Serious Managers Guide to AI Observability and Monitoring

From Pilots to Production:
Observing, Monitoring, and Governing Artificial Intelligence in the Enterprise

---

Jazper Carter

Cybersoft Publishing LLC

Fort Washington, MD 20744

Drclaude.net

First Edition April 2026

Contents

Introduction vii

1 The Blind Spot in Your AI Strategy..............................1

1.1 The Pilot-to-Production Gap 2
1.2 The Cost of Flying Blind 7
1.3 The Business Case for Observability Investment..12
1.4 Manager's Checklist 17
1.5 Takeaway 18

2 From DevOps Observability to AI Observability 19

2.1 The Three Pillars of Traditional Observability 20
2.2 Why AI Breaks the Rules of Traditional Monitoring 25
2.3 The AI Observability Maturity Spectrum 30
2.4 Manager's Checklist 35
2.5 Takeaway 36

3 Understanding What Can Go Wrong: Drift, Decay, and Degradation 37

3.1 Data Drift and Feature Drift 38
3.2 Concept Drift and Model Decay 43
3.3 Business Impact Patterns and Monitoring Priority.48
3.4 Manager's Checklist 52
3.5 Takeaway 53

4 The AI Observability Stack 55

4.1 Infrastructure and Data Pipeline Layers 56
4.2 Model Performance and Output Evaluation Layers 61
4.3 Open Standards and End-to-End Traceability.......66
4.4 Manager's Checklist 71
4.5 Takeaway 72

5 Monitoring Traditional ML Models in Production 73

5.1 Establishing Baselines and Choosing the Right Metrics 74
5.2 Drift Detection Methods and Alert Policies 79
5.3 The Retraining Decision Framework and Model Lifecycle Governance 83

5.4 Manager's Checklist 88
5.5 Takeaway 89
6 Monitoring LLMs and Generative AI in Production....91
6.1 Prompt Versioning, Token Economics, and the Five Dimensions of LLM Observability..........................92
6.2 LLM-as-a-Judge, Semantic Drift, and Automated Output Evaluation 97
6.3 Guardrails, Safety Monitoring, and Fallback Architectures 102
6.4 Manager's Checklist 107
6.5 Takeaway 108
7 Observing Agentic AI Systems 109
7.1 Decision Graphs, Execution Traces, and the Fundamental Observability Challenge 110
7.2 Tool Call Fidelity, Session Evaluation, and Multi-Agent Architectures 115
7.3 The Trust Gap: Governance, Human Oversight, and Closing the Autonomy-Observability Loop 120
7.4 Manager's Checklist 125
7.5 Takeaway 126
8 The Vendor Landscape: Choosing the Right Observability Platform 127
8.1 Market Segmentation and the Four Tiers of AI Observability Tooling..128
8.2 The Vendor Evaluation Framework: Twelve Criteria and a Build-vs-Buy Decision 133
8.3 Cost Modeling and Total Cost of Ownership at Scale 138
8.4 Manager's Checklist 142
8.5 Takeaway 143
9 Building Your Observability Architecture 145
9.1 Architecture Patterns That Scale.........................146
9.2 Data Flow, Storage, and the Observability Lake .150
9.3 Dashboards, Integrations, and Reference Architecture 155
9.4 Manager's Checklist 160
9.5 Takeaway 161

10 Alerting, Incident Response, and Escalation...........162
10.1Designing Alerts That Drive Action......................163
10.2The AI Incident Response Playbook 168
10.3Post-Incident Review and Organizational Learning 173
10.4Manager's Checklist 178
10.5Takeaway 178
11 Governance, Compliance, and Audit Readiness.....180
11.1The Regulatory Landscape for AI in Production..181
11.2Observability as Evidence and the Audit Trail.....186
11.3Building the Audit-Ready Governance Program .191
11.4Manager's Checklist 196
11.5Takeaway 197
12 Cost Optimization and FinOps for AI 198
12.1Understanding the AI Cost Structure...................199
12.2Optimization, Right-Sizing, and the FinOps Feedback Loop..204
12.3FinOps Frameworks and Vendor Negotiation for AI 208
12.4Manager's Checklist 213
12.5Takeaway 214
13 Building the AI Observability Team 216
13.1The Skills Gap and Why It Persists 217
13.2Role Definitions and Organizational Models 221
13.3Cross-Functional Collaboration and Upskilling....226
13.4Manager's Checklist 231
13.5Takeaway 232
14 The AI Observability Maturity Model 233
14.1The Five Maturity Levels Defined........................234
14.2Self-Assessment and Gap Analysis 239
14.3Roadmaps, Quick Wins, and Implementation Discipline 244
14.4Manager's Checklist 249
14.5Takeaway 249
15 The Road Ahead: Observability in the Age of Autonomous AI 251

15.1From Monitoring Models to Observing Decisions252
15.2Edge AI, Digital Twins, and Emerging Standards257
15.3The Manager's Strategic Mandate 262
15.4Manager's Checklist 267
15.5Takeaway 268
Closing...269

References 273

Introduction

Most enterprises do not know how their AI systems are actually performing in production. That is not an accusation; it is the observable result of a common and understandable pattern. Organizations invest heavily in building AI systems, in data preparation, model development, validation, and deployment, and then invest almost nothing in watching those systems once they are live. The model reaches the production inference endpoint, the project is declared a success, and the team moves on to the next initiative. What happens afterward, how the model behaves as the real world diverges from the training distribution, whether its outputs remain accurate and fair as conditions change, and whether its predictions still align with the business outcomes they were designed to drive, is largely unknown. That ignorance is expensive. It manifests as quiet revenue leakage, accumulated compliance exposure, and the kind of reputational incident that arrives without warning precisely because no one was watching. This book exists to change that pattern.

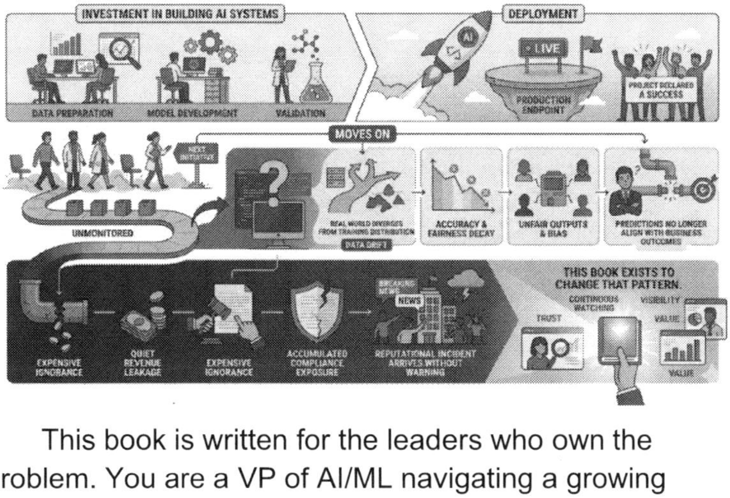

This book is written for the leaders who own the problem. You are a VP of AI/ML navigating a growing portfolio of production models with no unified visibility into their health. You are a data engineering manager whose pipelines feed AI systems but whose monitoring stops at the data layer. You are an ML platform lead who built the deployment infrastructure but lacks a systematic process for monitoring what the deployed models do. You are a CIO or enterprise architect evaluating observability investments and trying to understand what a mature AI monitoring program actually requires. You are an IT operations leader whose incident management processes were designed for deterministic software and now face the fundamentally different challenge of AI systems that fail quietly rather than visibly. This book provides the management framework, the technical literacy, and the organizational guidance that each of these roles needs to build, govern, and sustain a production AI observability program.

The book is organized into four arcs across fifteen chapters. The first arc, Chapters 1 through 4, establishes the foundational context: why AI systems require observability, how AI observability differs from traditional DevOps monitoring, what failure modes affect production AI systems, and what the architecture of a complete AI observability stack looks like. The second arc, Chapters 5 through 7, addresses the three major classes of production AI systems in detail: traditional predictive ML models, large language models and generative AI, and autonomous agentic systems, each with distinct failure modes and monitoring requirements. The third arc, Chapters 8 through 10, covers the operational disciplines of selecting and deploying observability technology, designing the supporting architecture, and building the alerting and incident response processes that translate observability signals into organizational action. The fourth arc, Chapters

11 through 15, addresses the governance, financial, organizational, and strategic dimensions of operating an enterprise AI observability program at scale, and looks ahead at the direction the discipline is heading.

Before proceeding, several terms used throughout this book deserve precise definitions because they are often used loosely in the industry. Observability is distinct from monitoring. Monitoring asks whether a predefined metric has crossed a predefined threshold. Observability asks whether you have enough information about a system's internal state to diagnose any failure mode, including ones you did not anticipate when you designed the monitoring rules. Drift is the generic term for changes in the statistical properties of an AI system's inputs or outputs over time; it encompasses data drift, meaning shifts in the distribution of raw input features; concept drift, meaning shifts in the relationship between input features and target outcomes; and feature drift, meaning changes in engineered feature values caused by pipeline modifications or upstream data source changes. Model decay is the broader phenomenon in which a model's predictive capability declines over time as the cumulative effect of these drift mechanisms accumulates. Guardrails are the safety and policy controls that constrain the outputs of generative AI systems; guardrails monitoring tracks whether those controls are functioning correctly and whether the rate of policy violations is stable or trending. A trace, in the OpenTelemetry sense, is a structured record of a request's path through a distributed system; each segment of that path is a span, and the collection of spans forms the context graph that enables end-to-end root cause analysis. Outcome observability is the emerging practice of monitoring not just whether the model's outputs are statistically well-formed, but also whether the real-world decisions and actions those outputs drive produce the intended results.

By the end of this book, you will be equipped to perform a structured self-assessment of your organization's current AI observability maturity and identify the specific capability gaps that represent the highest risk to your production AI portfolio. You will be able to design a monitoring architecture that covers the infrastructure, data pipeline, model performance, output evaluation, and agent observability layers with appropriate tooling at each layer. You will be able to build a business case for observability investment using the risk-quantification framework presented in Chapter 1 and the cost-modeling approaches in Chapter 12. You will be able to evaluate vendor platforms against a structured scorecard, design the governance artifacts required by regulatory frameworks, and build the organizational structure, roles, and review cadences that make an observability program sustainable. You will be able to lead the conversation with your engineering teams, your compliance function, and your executive stakeholders about what it actually means to take production AI seriously.

The manager's role in AI observability is not to write the code or configure the dashboards. It is to set the requirements, allocate the resources, own the accountability structure, and ensure that the feedback loops that observability creates are actually connected to the decisions that matter. Every AI system your organization deploys is a commitment: a commitment to your customers that the system will behave correctly, to your regulators that it will operate within bounds, and to your organization that it will deliver the business value that justified the investment. Observability is the operational infrastructure that keeps those commitments honest. This book will show you how to build it.

1 The Blind Spot in Your AI Strategy

A large regional insurer spent 18 months and $4 million building a claims-triage model that routes property claims to the correct adjuster tier. The model passed every validation gate on the ML platform, cleared the model risk committee review, and went live on the production inference endpoint in March. By July, the VP of Claims Operations noticed adjuster productivity falling, and average cycle time rising by 11%. No alert had fired. The feature store pipeline had silently shifted to a key vehicle-age feature after an upstream data vendor changed its schema in April. The model was still running, but it was routing claims incorrectly. The damage was real: delayed payouts, customer complaints, and a reinsurance partner flagging anomalies in loss ratios. The team had no observability tooling on the model's output distribution, no drift threshold, and no process for reviewing model health between quarterly cycles.

Most enterprises treat AI deployment as a project milestone rather than an operational commitment. Once a model reaches the production inference endpoint, the project team disperses, and the model runs without systematic observation. Production AI systems operate in open-world environments where the data in month seven rarely matches the training data from month one. Without observability, the gap between model behavior and business expectation widens in silence. Senior managers who do not establish AI observability programs before deployment are accumulating invisible technical debt that surfaces as operational failure.

The business stakes are substantial. A credit-scoring model that has drifted can create fair-lending exposure that dwarfs the cost of any monitoring platform. An LLM-powered customer service agent that degrades burns customer trust faster than the original deployment built it. Forrester Research estimates that organizations with mature AI observability programs realize a 357 percent return on investment over three years, primarily through reduced incident costs and faster remediation. Organizations without observability absorb those costs invisibly as lost revenue, remediation labor, and compliance penalties.

1.1 The Pilot-to-Production Gap

The distance between a successful AI pilot and a trustworthy production system is not measured in code commits or model accuracy scores. The operational disciplines an organization has built to support a system that must perform reliably every day against real-world data have never been seen. Most enterprises close the deployment gap but leave the observability gap entirely open, and that is where models go quietly wrong.

Why Models That Work in Testing Fail Silently in Production

A model trained on historical data is optimized for a world that no longer exists. The training distribution captures a snapshot of customer behavior, market conditions, or sensor readings as they were during the data collection window. Production data arrives from a continuously evolving world. Seasonal shifts, product changes, and competitor actions alter the statistical properties of incoming data in ways no static validation set anticipates. When those shifts exceed the model's tolerance, their outputs degrade, but the inference endpoint keeps serving predictions with the same apparent confidence.

The silence of failure is the central danger. A traditional software bug generates an exception, a stack trace, or a failed health check. A model that has drifted generates nothing of the sort. It produces outputs that are syntactically valid, within expected data type ranges, and returned at normal latency. The only signal that something is wrong lives in the semantic relationship between the model's outputs and the business outcomes those outputs are supposed to drive. Without a monitoring layer that tracks that relationship, the failure propagates undetected through every downstream decision the model informs.

The pilot-to-production pathway exposes this problem sharply. Pilot conditions are carefully controlled: curated data, limited scope, and close supervision from the data science team. Those conditions vanish at the production scale. Alert rules, if they exist at all, check latency and error rates, not output distribution or business metric alignment. Managers who have not mandated an observability plan before a model goes live have handed the autopilot controls to a system with no attitude indicator.

Diagram 1.1 — Pilot vs Production Observability Gap

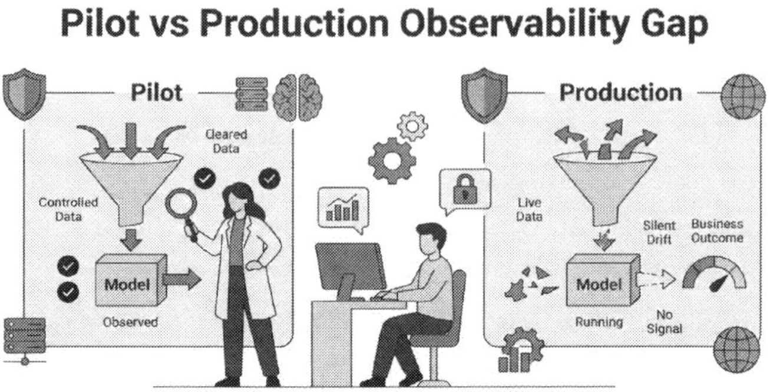

What Observability Actually Means for AI Systems

Observability is not the same as monitoring. Monitoring asks whether a predefined metric has crossed a predefined threshold. Observability asks whether you have enough information about a system's internal state to diagnose any failure mode, including failure modes you did not anticipate when you wrote the alert rules. In traditional software engineering, that distinction is the foundation of modern site reliability practice. In AI systems, the distinction is even more consequential because the failure modes are probabilistic, emergent, and often invisible to infrastructure-level tooling.

For an AI system, observability requires three capabilities that traditional application performance management platforms do not provide. First, the ability to capture model inputs, outputs, and eventual ground-truth outcomes linked across time. Second, statistical methods to detect when the input or output distribution has shifted beyond acceptable bounds. Third, evaluation frameworks that assess whether model outputs are correct, safe, fair, and aligned with the intended use case, not just whether they arrived on time.

This broader definition of observability has direct organizational implications. An observability program for AI is not a single tool purchase. It is a set of practices, data contracts, evaluation cadences, and escalation paths spanning the ML platform team, the data engineering team, the AI SRE function, and the governance office. Managers who conflate observability with uptime monitoring will fund the wrong capabilities and leave the most consequential risks unaddressed.

Diagram 1.2 — Monitoring vs Observability Comparison

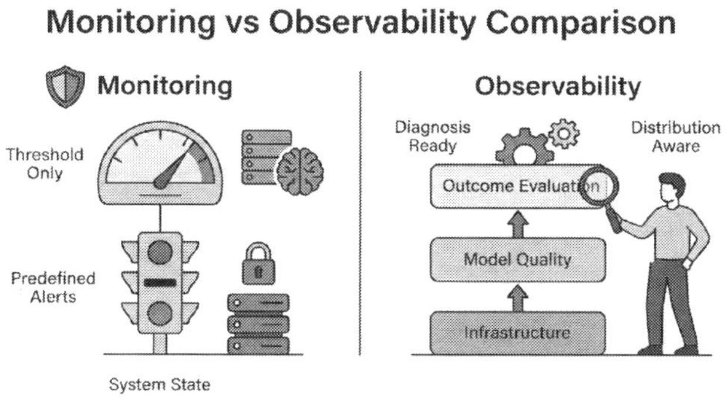

Organizational Conditions That Create Blind Spots

Blind spots in AI production are not purely technical. They are organizational. Three structural conditions reliably produce unobserved AI systems. The first is ownership ambiguity: the data science team built the model, the data engineering team manages the pipeline, the platform team runs the inference endpoint, and no one has explicit accountability for monitoring the model's behavior in production. Each team assumes another has it covered. None of them does. Without a named observability lead and a defined RACI for AI production health, coverage gaps are the default outcome.

The second condition is stage-gate thinking. Many enterprises treat AI as a project that ends at deployment. The model passes validation, is cleared by the model risk committee, and is promoted to production. At that point, the project is marked complete, and resources are reallocated. The model enters a maintenance posture with no dedicated monitoring cadence and no retraining trigger. The third condition is metric mismatch: the metrics tracked after deployment, typically latency and error rate, are what existing infrastructure tooling already produces. The metrics that actually matter for AI system health require purpose-built instrumentation that most organizations have not invested in.

Addressing these organizational conditions requires management action, not technology purchases. The ML platform lead must define a model production health process before the first model reaches production. The VP of AI/ML must establish a cross-functional review cadence that covers data quality, model performance, and business outcomes. The observability lead must own a documented escalation path so that when a drift threshold fires, a named individual decides to retrain, roll back, or escalate.

Diagram 1.3 — Three Structural Blind Spot Conditions

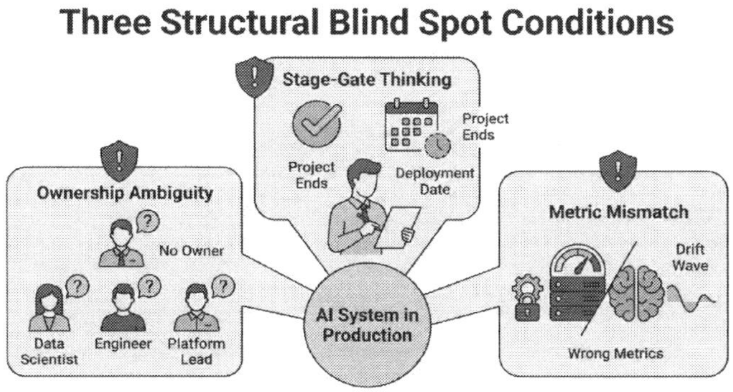

1.2 The Cost of Flying Blind

Unobserved AI systems do not fail dramatically. They erode quietly, and the costs accumulate across three dimensions: operational efficiency losses, compliance exposure, and reputational damage. Each dimension is quantifiable, and each represents a risk category that senior managers are already accountable for in adjacent domains. The discipline of AI observability extends existing risk management practice into the AI layer, where it is currently absent in most enterprise environments.

Revenue Leakage from Silent Model Degradation

Every production AI system drives or informs a business process. A demand forecasting model shapes inventory orders. A propensity-to-churn model triggers retention outreach. A credit-underwriting model determines which loan applications advance. When any of these models drift, the downstream business process operates on systematically wrong information. The revenue impact is not hypothetical; it is the difference between what the model would have recommended at peak calibration and what it is actually recommending after drift has accumulated. In high-volume systems, that gap compounds across millions of decisions before anyone notices.

The insidious characteristic of revenue leakage due to model drift is that it rarely manifests as a discrete event. Instead, it shows up as a slight decline in conversion rate, a modest uptick in return rates, or a gradual increase in claim frequency that finance attributes to seasonal variation. Because the signal is diffused and the attribution to model quality is not obvious, organizations often bear the loss for quarters before connecting it to the AI system. ML platform teams with output distribution monitoring and automated alerting catch these events within days, not quarters.

Quantifying the risk requires an exercise most organizations have not performed: estimating the revenue sensitivity of each production AI model to a ten percent accuracy degradation. For a recommendation engine driving e-commerce conversion, a 10% decline in accuracy can translate into a seven-figure annual revenue reduction. Building those sensitivity estimates before deployment and using them to set alert thresholds transforms observability from an infrastructure cost into a revenue-protection investment.

Diagram 1.4 — Revenue Leakage from Drift Accumulation

REVENUE LEAKAGE FROM DRIFT ACCUMULATION

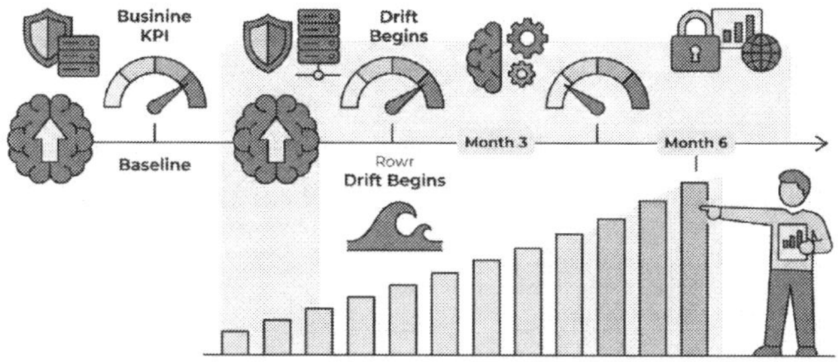

Compliance Exposure and Regulatory Risk

Regulators in financial services, healthcare, and increasingly across all industries are treating production AI systems as regulated assets. SR 11-7 requires banks to maintain evidence of model performance monitoring, including documentation of thresholds, trigger events, and corrective actions. The EU AI Act extends similar requirements to high-risk AI applications. HIPAA enforcement actions have identified AI-assisted diagnostic tools as covered systems that require ongoing monitoring for bias and accuracy. Managers without an observability program cannot produce the audit artifacts these frameworks require.

The compliance exposure from unobserved AI compounds when the model's degradation has a disparate impact dimension. A credit-scoring model that has drifted in a way that disadvantages protected classes, not through deliberate design but through upstream data shift, creates fair-lending liability that is difficult to defend without a continuous monitoring record. Regulators ask not only whether the model was fair at validation time but whether the organization had a system in place to detect post-deployment fairness drift. An observability program that tracks output distributions across demographic segments and alerts when gaps widen provides exactly the evidence trail a defensible compliance posture requires.

The business case for compliance-aligned observability is straightforward: the cost of a purpose-built monitoring platform is a fraction of the cost of a single regulatory enforcement action, consent order, or class-action settlement. Managers should frame observability investment not as an optional technology spend but as a risk mitigation mechanism that belongs in the same budget conversation as cyber insurance, internal audit, and model risk management staffing. Responsible by design means that the governance infrastructure is built before the regulators come asking.

Diagram 1.5 — Compliance Risk from Unobserved AI

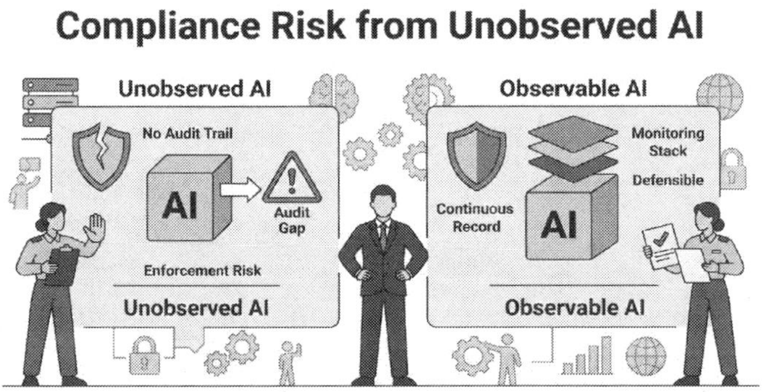

Reputational Damage and the Trust Erosion Cycle

Enterprise AI systems operate in contexts where trust is both an asset and a liability. Customers trust that a bank's AI-driven service recommendations are in their interest. Employees trust that an AI-assisted HR screening tool treats applicants fairly. Clinicians trust that an AI-augmented diagnostic support tool flags the right cases. When any of these systems produces a visible failure, the reputational damage is not proportional to the technical severity of the drift event. It is proportional to the public perception of whether the organization knew and did nothing.

The erosion of trust follows a predictable pattern. A model degrades in production. A user or journalist encounters an anomalous output and investigates. The organization has no clear monitoring record to demonstrate that it was actively tracking system health. The narrative that emerges is that the organization deployed AI carelessly and left it unsupervised. That narrative is far more damaging than the underlying technical incident because it speaks to the organization's character rather than to an isolated software failure. Rebuilding trust after a high-profile AI failure can cost multiples of the original deployment investment in PR, regulatory engagement, and product redesign.

The management response to this risk is an observability program that creates a visible, documented record of ongoing stewardship. When a model failure does occur, and over a long enough time horizon, some failure is inevitable in any complex production system. The organization's ability to produce an incident timeline, a root cause analysis grounded in monitoring data, and a corrective action plan is the difference between a contained technical incident and a reputational crisis. Observability is not only about catching failures early; it is about demonstrating to every stakeholder that the organization takes its AI systems seriously enough to monitor them continuously.

Diagram 1.6 — AI Trust Erosion Cycle

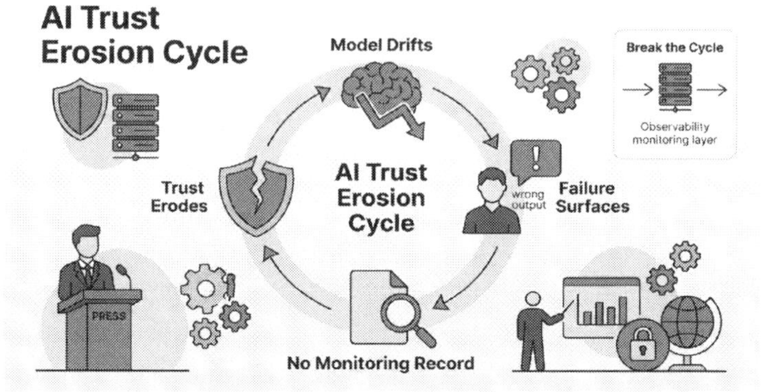

1.3 The Business Case for Observability Investment

Building an AI observability program requires investment in budget, staffing, and tooling. Senior managers who have not previously funded observability need a concrete financial frame to justify that investment. The case rests on three pillars: quantified risk reduction, demonstrated return from industry data, and the compounding value of observability as the AI portfolio grows. Each pillar translates into the budget language that finance leaders and executive sponsors respond to.

The 357 Percent ROI Finding and What It Measures

Forrester Research's Total Economic Impact study on AI observability platforms found that organizations with mature monitoring programs realized a 357 percent return on investment over three years. That figure is not a marketing number; it is the result of a structured economic model that accounts for avoided incident costs, reduced mean time to resolution for model failures, labor savings from automated drift detection, and value recaptured from models that would otherwise have been retired prematurely due to undiagnosed performance degradation. Each category represents a real cost that organizations without observability absorb without recognizing it.

The avoided incident cost component is typically the largest single driver of ROI. An undetected drift event in a high-volume production model not only incurs the labor hours required to diagnose and fix it after discovery, but also the cumulative business impact of every wrong decision the model made during the period between drift onset and detection. In a study of financial services deployments, the average gap between drift onset and detection in organizations without automated monitoring was 73 days. At the transaction volumes typical of retail banking AI systems, 73 days of degraded performance represents a quantifiable revenue and risk exposure that dwarfs the cost of any observability platform.

Managers building a business case should construct their own version of the ROI model. Start with revenue sensitivity analysis for each production model: estimated business impact per percentage point of accuracy degradation, and how many days the current process takes to detect a ten-percentage-point drift event. Multiply those numbers together for each model, sum them across the portfolio, and compare the result to the all-in cost of the observability program. In most enterprise AI portfolios of five or more production models, that calculation produces a compelling positive result before accounting for compliance risk avoidance.

Diagram 1.7 — Observability ROI Value Drivers

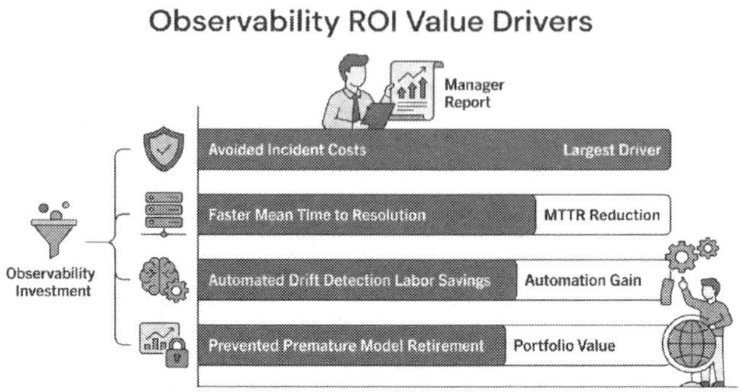

Observability as Portfolio Infrastructure

A single-model justification for observability investment significantly understates its value. The correct unit of analysis is the AI portfolio, because observability infrastructure is a shared platform that generates compounding returns as the number of production models grows. The first model's monitoring setup requires a full fixed-cost investment in instrumentation, data pipelines, alerting infrastructure, and review-cadence design. The second model's monitoring setup reuses that infrastructure and costs a fraction of the first. By the time an organization has ten production models, the per-model marginal cost of observability is a small fraction of the initial investment. At the same time, the aggregate risk exposure it covers has grown tenfold.

This portfolio framing also changes how managers should think about investment timing. The instinct to wait until the portfolio is large enough to justify the platform cost is wrong for two reasons. First, the risk of undetected failure is highest for early production models, precisely because there is no institutional muscle for monitoring. Second, building observability after models are already running requires retrofitting instrumentation into systems not designed to emit the required data. Building observability-ready systems from the start, using standardized telemetry with OpenTelemetry conventions, is far less expensive than adding observability to live production systems.

The ML platform lead's role is to establish the shared observability infrastructure as a foundational platform service, not as a feature added to individual model projects. When observability is a platform service, it is governed by platform standards, funded from the platform budget, and maintained by platform engineers rather than individual model teams. That organizational structure ensures that every model entering production inherits a monitoring baseline, without requiring each project team to design its own. Mission-aligned observability means the platform is built to support the organization's long-term AI ambitions, not just the model in front of it today.

Diagram 1.8 — Observability Platform Portfolio Scale Effect

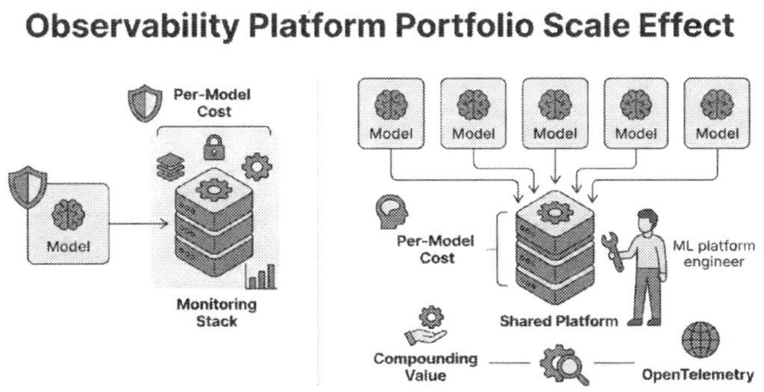

Building the Internal Funding Argument

Executive sponsors and finance leaders respond to investment proposals that quantify risk. The internal funding argument for an AI observability program must be framed in the language of risk management. The conversation is not about dashboards and drift metrics; it is about the probability and magnitude of business losses the observability program will prevent. That reframe requires managers to do the analytical homework of estimating those losses before the budget meeting.

A practical framing follows three steps. First, inventory every production AI system and classify each by business impact category: revenue-driving, cost-reducing, risk-managing, or compliance-required. Second, for each system, estimate the maximum tolerable degradation window, the estimated daily exposure to undetected degradation, and the current mean time to detect a significant performance issue. Third, express the total unmitigated risk exposure as an annualized dollar figure. That figure becomes the numerator in the return-on-investment calculation.

Managers should build a roadmap showing how investment scales with the portfolio. A phased plan that starts with automated monitoring for the two or three highest-impact production models, establishes shared platform infrastructure in the first six months, and extends coverage to the full portfolio over eighteen months is more defensible than a single large capital request. Observability programs that grow incrementally demonstrate fiscal discipline while building the organizational capability that complex, autonomous AI systems will eventually require. The alternative to investment is not zero cost; it is the unbounded, unquantified cost of operating AI systems that no one is watching.

Diagram 1.9 — Internal Observability Funding Argument

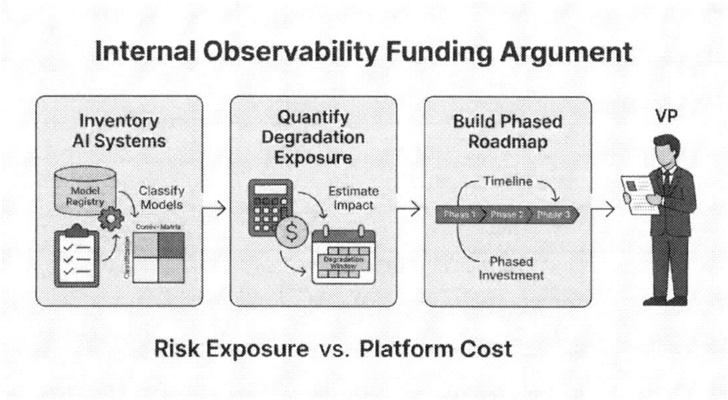

1.4 Manager's Checklist

- Before any model reaches production, document the estimated revenue or risk exposure per day of undetected performance degradation.
- Assign a named observability lead with explicit accountability for each production AI system's monitoring health.
- Require a baseline output distribution report for every model at the time of production promotion, signed by the ML platform lead.
- Establish a RACI matrix for AI production incidents that covers detection, diagnosis, escalation, and corrective action.
- Set a maximum tolerable detection window for each production model class and document it as an operational SLA.
- Budget observability tooling as shared platform infrastructure, not as a line item in individual model project budgets.
- Schedule a quarterly AI portfolio health review covering drift status, alert history, and business KPI alignment for every production model.

- Require all new AI systems to emit standardized telemetry using OpenTelemetry conventions before receiving production approval.

1.5 Takeaway

The blind spot in most enterprise AI strategies is not a gap in technical capability; it is a gap in operational discipline. Organizations that have invested heavily in building and deploying AI have rarely invested proportionally in watching those systems perform in the real world. The result is a growing population of production models that accumulate invisible drift, exposing their organizations to entirely preventable revenue, compliance, and reputational risks.

AI observability is the operational discipline that closes that gap: continuous, structured visibility into the behavior of production AI systems so that failures are detected before they become crises, compliance evidence exists before regulators ask, and business stakeholders can trust the AI informing their decisions. Building that discipline is a management responsibility.

2 From DevOps Observability to AI Observability

A global logistics company had built a mature DevOps observability practice. Every microservice emitted structured logs, traces ran through Datadog, and SREs carried well-tested runbooks for every alert tier. When the platform team deployed a shipment-delay prediction model, they assumed the existing stack would cover it. Fourteen weeks later, the model's prediction accuracy declined eighteen percentage points due to feature drift in carrier transit-time data. Datadog had captured inference latency perfectly and never alerted, because the latency was fine. The model answered faster than ever, but wrong. The engineering team had monitored the container and missed the model entirely. Conflating application observability with AI observability had cost the operations team a quarter of the manual exception-handling labor the model was supposed to eliminate.

Every enterprise IT organization has made substantial investments in APM and DevOps observability. Those investments are real and valuable, but they address a fundamentally different class of system. Traditional software is deterministic: given the same input, it produces the same outputs. Infrastructure-layer tooling can tell you whether the system is running, how quickly it responds, and whether it is generating errors. AI systems require a different scope because what can go wrong is not the code; it is the model's reasoning, and that reasoning is invisible to infrastructure-layer tools.

The organizational risk of conflating DevOps observability with AI observability is a false sense of coverage that is more dangerous than acknowledged ignorance. Engineering teams with mature APM practices underestimate the AI-specific monitoring gap. Budget conversations are foreclosed, specialized tooling is not purchased, and the model ships into the same stack as the API it runs behind. Managers who understand the precise distinction between what APM tools cover and what AI observability requires can allocate investment and assign accountability accurately.

2.1 The Three Pillars of Traditional Observability

Traditional observability in DevOps practice is built on three signal types: logs, metrics, and traces. Understanding what those pillars can and cannot tell you about a system is the foundation for understanding where AI observability extends beyond them. The three pillars are powerful for answering infrastructure questions; they are structurally incapable of answering model quality questions, and that limitation is not a vendor gap but a conceptual one.

Logs, Metrics, and Traces in Production Systems

Logs are timestamped records of discrete events: a request received, a function called, an exception thrown. In a well-instrumented application, they capture the narrative of what the system did and when it did it. They are invaluable for debugging discrete failures, such as bad configuration, a database timeout, or a null value in a field. Modern logging platforms built on OpenTelemetry's log signal structure these records for machine-readable querying and integrate with alerting pipelines. For traditional software, a comprehensive log strategy provides the evidentiary record needed to diagnose nearly any failure.

Metrics are time-series measurements: request rate, error rate, latency percentiles, CPU utilization, and queue depth. They power operational dashboards and threshold-based alerting that SREs monitor. Tools like Grafana, Datadog, New Relic, and Dynatrace are built primarily around metric collection, visualization, and alerting. Metrics tell operators whether the system is under load, whether it is responding within SLA, and whether infrastructure resources are consumed at expected rates. They are excellent at answering whether the system is running.

Traces record a request's path through a distributed system: which services it touched, how long each hop took, and where latency spikes or errors occurred. OpenTelemetry's distributed tracing standard enables engineers to follow a single user request through dozens of microservices and identify the exact span where something went wrong. Traces are powerful for diagnosing latency issues and understanding request flow in complex architectures. For an AI system, a trace tells you that the inference endpoint responded in 120 milliseconds; it does not tell you whether the inference was correct.

Diagram 2.1 — Three Traditional Observability Pillars

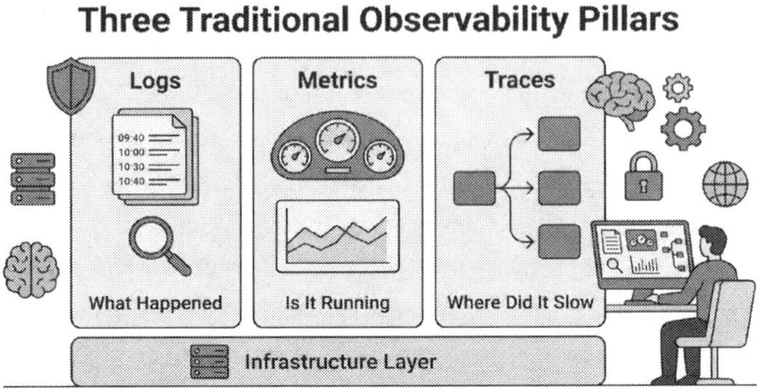

What the Three Pillars Cannot Measure

The three pillars of traditional observability share a common architectural assumption: the system under observation is deterministic. If the correct inputs arrive and the code executes without error, the output is correct by definition. That assumption is foundational to the design of APM tools. Threshold alerts fire on error rates and latency spikes because those signals reliably indicate incorrect behavior in deterministic software. There is no equivalent signal for an AI model that is executing without error but producing outputs that are systematically wrong.

The gap is precise and consequential. A model serving predictions through a REST endpoint can have zero errors, sub-100-millisecond latency, and perfect infrastructure health metrics, yet its predictions are drifted 18 percentage points from validation accuracy. Every log entry looks clean. Every metric is green. Every trace completes successfully. The only evidence that the model is underperforming lies in the statistical relationship between its output distribution today and its distribution at training time, which is not captured by logs, metrics, or traces.

Understanding this gap operationally means recognizing that 'Is it running?' is necessary but not sufficient for AI systems. Running without errors is the minimum condition for a software service. For an AI model, the sufficient condition is: are its outputs still aligned with the use case it was deployed to serve? Answering that question requires a fourth signal type that is not present in the traditional stack: model-quality signals covering output distributions, prediction-confidence calibration, feature-importance stability, and ground-truth outcome comparisons.

Diagram 2.2 — APM Coverage Gap for AI Systems

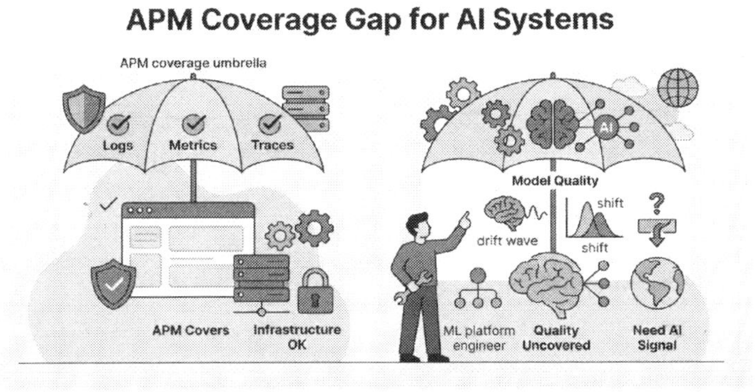

Extending the Pillars: The AI-Specific Signals

AI observability does not discard the three traditional pillars; it extends them with signal types designed specifically for model behavior. AI-specific signals include feature distributions, prediction distributions, confidence calibration metrics, and outcome feedback. Feature distributions capture statistical properties of model input variables over time. Prediction distributions track whether model output values have shifted. Confidence calibration measures whether the model's stated confidence scores accurately reflect its actual accuracy. Outcome feedback compares model predictions against eventual ground-truth labels.

Tools in the AI-native observability category, including Arize AI, Fiddler AI, and Evidently AI, were designed from the ground up to collect and analyze these signal types. They instrument the model serving layer to capture inputs and outputs at inference time, compute statistical drift metrics such as the Population Stability Index and Jensen-Shannon divergence on rolling windows, and generate alerts when distribution shift exceeds configurable thresholds. This is the observability work that Datadog's AI module and New Relic's ML observability extension are progressively adding, though AI-native tools typically offer deeper statistical sophistication.

For ML platform engineers, the practical integration challenge is connecting the AI signal layer to existing APM infrastructure. OpenTelemetry's GenAI semantic conventions, under active CNCF development, provide a standardized way to emit AI-specific spans alongside traditional application traces. When an inference request flows through an OpenTelemetry-instrumented pipeline, the resulting trace can carry the latency span from the API gateway and model-specific attributes, including input feature values, the model's output prediction, and the confidence score. That unified trace structure is the architectural foundation for AI observability, integrating with rather than replacing the existing DevOps investment.

Diagram 2.3 — Extended AI Observability Signal Stack

Extended AI Observability Signal Stack

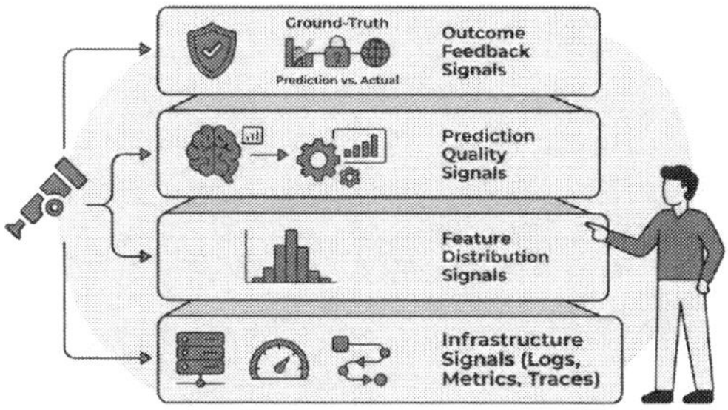

2.2 Why AI Breaks the Rules of Traditional Monitoring

AI systems violate the architectural assumptions of traditional monitoring in three specific ways: non-determinism, probabilistic reasoning, and emergent behavior. Each violation creates a monitoring blind spot that no amount of infrastructure tuning can close. Understanding why each violation occurs, and what it means for a monitoring program, is the conceptual prerequisite for designing an AI observability approach that actually works.

Non-Determinism and Probabilistic Outputs

A traditional software function, given the same inputs, always returns the same output. That determinism is not incidental; it is a design property that enables testing, reproducibility, and defect isolation. When a bug appears, engineers can reproduce it by replaying the same input and observing the same wrong output. Traditional monitoring implicitly relies on this property. An alert that fires when the error rate exceeds five percent is meaningful because errors are discrete, reproducible events. The presence or absence of an error is unambiguous.

Machine learning models are probabilistic systems. A classification model outputs a probability distribution over class labels; the predicted class is the highest-probability label, but the model is not certain in the way a conditional statement is. A generative AI model, given the same prompt, may produce different outputs on consecutive calls because the sampling process involves stochastic elements. Neither behavior is a bug; both are inherent properties of the model. The monitoring implication is that correctness cannot be evaluated on a per-request basis; it must be evaluated statistically across a population of requests over time.

This shift from per-request correctness checking to population-level statistical evaluation requires a different monitoring architecture. Instead of an alert that fires when a specific request produces a wrong answer, AI observability uses rolling statistical windows to track whether the output distribution remains consistent with the validation-time baseline. Tools like Evidently, Arize, and the monitoring modules in Databricks MLflow implement this pattern. The ML platform engineer defines a reference distribution at deployment time and configures statistical tests to trigger when live data diverges from the specified threshold.

Diagram 2.4 — Deterministic vs Probabilistic System Monitoring

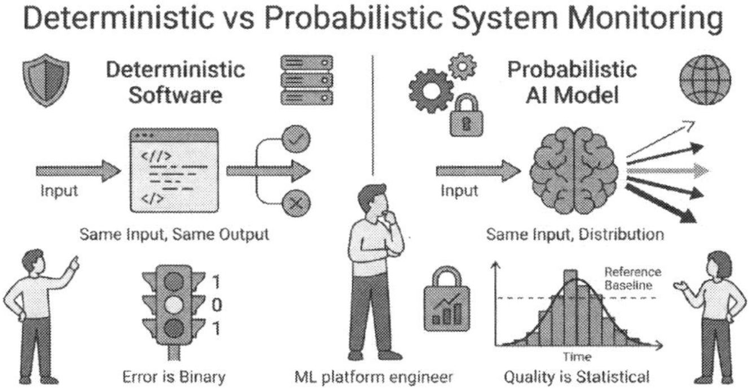

Emergent Behavior in Large Models

Large language models and deep learning systems exhibit emergent behaviors: failure modes do not present at smaller scales and not explicitly programmed. An LLM may produce confident, fluent, and entirely fabricated responses to questions outside its training distribution. A computer vision model trained on specific medical imaging equipment may fail silently on a different manufacturer's scanner due to subtle pixel-level distribution differences. These behaviors do not manifest as errors in any traditional sense; they manifest as qualitatively wrong outputs that the infrastructure layer cannot distinguish from correct ones.

Emergent failure modes are particularly challenging for organizations that rely on pre-deployment testing to certify model behavior. Evaluation on a held-out test set measures performance on data drawn from the training distribution. It provides no coverage for failure modes that emerge only under real-world distribution shifts, adversarial inputs, or scale effects not present in the test environment. This means that a model with a clean validation scorecard can still exhibit emergent failure modes in production that the evaluation process never surfaced.

The operational response to emergent behavior is a monitoring program that tracks output semantics rather than only output statistics. For traditional ML models, statistical drift detection is often sufficient. For LLMs and generative systems, the monitoring program must include evaluation of quality dimensions not captured by numerical distributions, such as factual accuracy, coherence, policy compliance, and semantic relevance. Tools like Langfuse, LangSmith, and Arize Phoenix continuously run automated quality checks on sampled outputs, enabling early detection of emergent degradation before it becomes visible as business impact.

Diagram 2.5 — Emergent Failure Detection Architecture

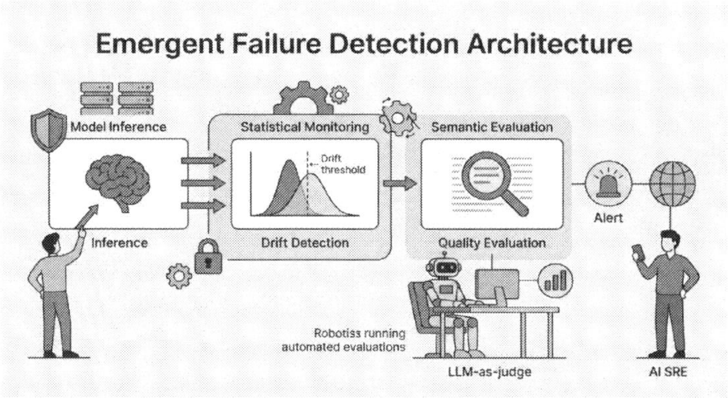

The Is It Running vs Is It Right Distinction

The clearest way to frame the gap between DevOps observability and AI observability is through two questions that demand entirely different monitoring architectures. 'Is it running?' is an infrastructure question: is the service alive, responding within SLA, and generating no errors? APM platforms answer this well. 'Is it right?' is a model quality question: are outputs still aligned with the intended use case, and are predictions still accurate relative to the business outcomes they drive? No APM platform answers this question without AI-specific extensions.

The operational implication is that an AI system can be unambiguously running while simultaneously being critically wrong. Both conditions coexist indefinitely because infrastructure health signals and model quality signals are measured at entirely different levels of the system stack. An SRE monitoring the Datadog dashboard for an inference service and seeing green status on all checks is providing accurate and complete coverage of the infrastructure layer. That coverage says nothing about whether the model running inside that infrastructure is still calibrated, unbiased, and aligned with its intended function.

Managers who understand this distinction will structure their AI observability programs with two distinct coverage areas, each with its own tooling, ownership, and alerting strategy. Infrastructure observability remains the SRE team's responsibility using existing APM platforms. Model quality observability becomes the ML platform team's responsibility using purpose-built AI monitoring tools. The two coverage areas are complementary, not redundant; an incident triggering in the AI observability layer may require SRE action at the infrastructure layer, so workflows must be integrated even as tooling and accountabilities remain distinct.

Diagram 2.6 — Two Observability Coverage Areas

Two Observability Coverage Areas

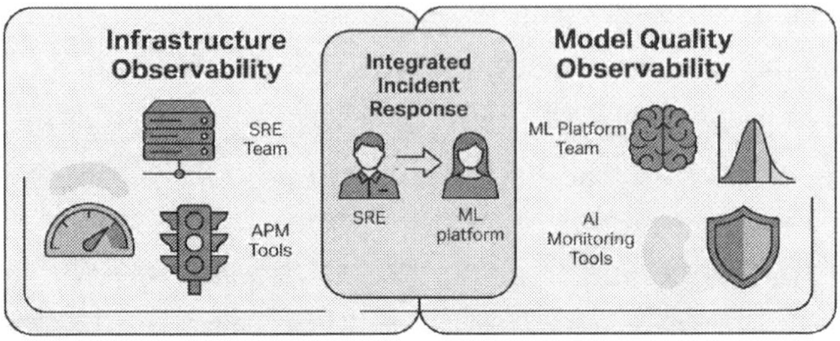

2.3 The AI Observability Maturity Spectrum

Organizations do not achieve AI observability in a single step. They move along a maturity spectrum from reactive ignorance to predictive intelligence, and each level corresponds to specific capabilities, tooling investments, and organizational behaviors. Understanding where an organization sits on that spectrum is the starting point for a prioritized investment roadmap. Managers who can accurately self-assess their observability maturity can build a sequenced improvement plan rather than attempting advanced capabilities before foundational ones are in place.

Reactive Alerting: The Starting Point

At the reactive level of AI observability maturity, organizations discover model failures when end users or business stakeholders report anomalous behavior. There is no proactive monitoring layer; the feedback mechanism is customer complaints, analyst exceptions, or a sudden drop in a business KPI that someone eventually traces back to the AI system. This is the operational posture of most enterprise AI programs in their first two to three years, not because managers deliberately chose it, but because the deployment investment consumed all available attention and budget, leaving nothing for post-deployment monitoring.

Reactive detection has a predictable cost structure. The mean time to detect a model failure is measured in weeks or months rather than hours, and each day of undetected failure represents accumulated business impact: wrong decisions made, wrong outputs delivered, wrong actions taken downstream. The remediation process at the reactive level is expensive because it begins with root-cause analysis from scratch, often requiring the data science team to rebuild context for a model they have not touched since deployment. Post-incident retooling invariably surfaces the monitoring infrastructure that should have been built before deployment.

Moving from reactive to the next maturity level requires two specific capabilities: a system to capture and store model inputs and outputs at inference time, and a scheduled process to review the output distribution against the validation baseline. The first is a data engineering task: instrument the inference endpoint to write each prediction to a feature store or data warehouse table. The second is an analytical task: compute basic statistical drift metrics weekly using Evidently or Great Expectations. Those two capabilities alone move an organization from reactive to basic observability and provide the foundation for more sophisticated monitoring.

Diagram 2.7 — Reactive AI Observability Pattern

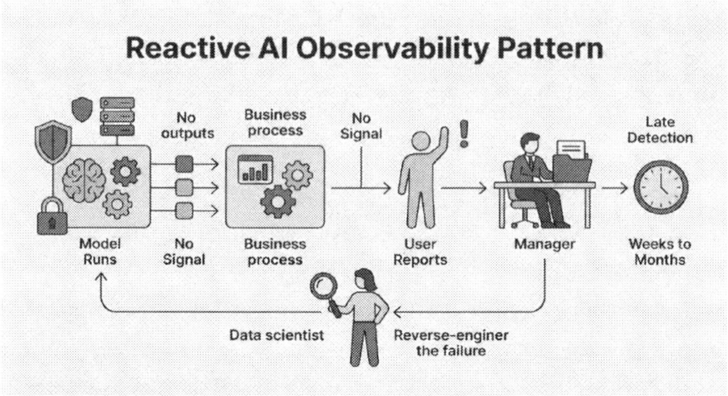

Proactive Monitoring and Automated Detection

At the proactive maturity level, the organization has deployed automated drift detection that runs continuously against production model outputs. Alerts fire based on statistical thresholds rather than user complaints. The ML platform team has defined a reference distribution for each production model, established alert tiers, and assigned a named on-call owner to receive and triage drift alerts. This level typically requires a purpose-built AI monitoring tool; Arize AI, Fiddler AI, or the Evidently open-source framework are common choices, often integrated alongside Datadog or Grafana dashboards to provide a unified operational view.

The threshold-setting process is a business judgment, not a statistical one. The ML platform lead must work with each model's business owner to establish what level of distribution shift constitutes an actionable warning versus an acceptable variation. A fraud detection model requires a tight threshold because even a small degradation in accuracy can carry significant financial exposure. A demand forecasting model for a stable product may tolerate wider drift because small errors wash out at aggregate levels. Thresholds calibrated to business tolerances produce actionable alerts; thresholds set by statistical convention alone generate alert fatigue.

The proactive investment in tooling has a clear, quantifiable payback. An automated detection system that reduces mean time to detect a significant drift event from eight weeks to four hours eliminates roughly 55 days of accumulated business impact per incident. At any reasonable estimate of the per-day cost of a degraded production model, the platform cost is recovered within a small number of avoided incidents. Managers building the business case should document the mean time to detect for their most recent model failure incidents and use those actual figures rather than theoretical estimates.

Diagram 2.8 — Proactive AI Monitoring Alert Flow

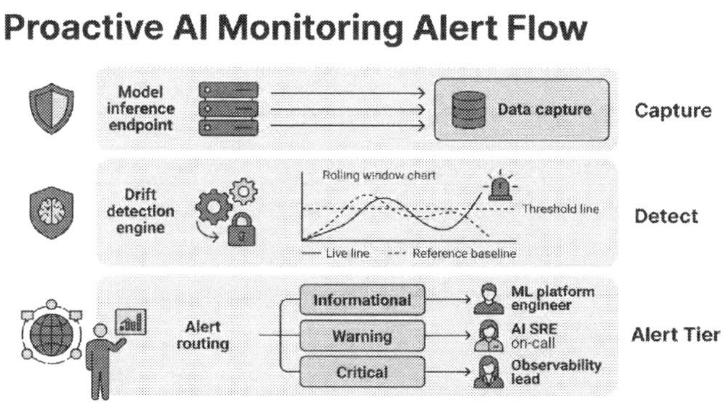

Predictive Intelligence: The Target State

At the predictive intelligence level, the observability program shifts from detecting degradation after it occurs to anticipating it before the alert threshold is reached. This capability adds trend analysis, upstream data pipeline monitoring, and model-specific leading indicators that correlate with future performance degradation. When upstream feature distributions begin to shift in patterns that historically precede accuracy decline, the system triggers an early warning, allowing the ML platform team to investigate before the model's output quality degrades.

Reaching the predictive level requires a longitudinal history of observability data. An organization needs at least six to twelve months of continuous monitoring data, several documented drift events with their upstream precursors, and enough model incidents to build a pattern library. That history enables the ML platform team to identify which upstream signals reliably predict downstream degradation for each model class. Some models degrade when a specific feature's variance increases; others show leading indicators in pipeline freshness metrics before output quality drops. The pattern is model-specific and emerges from operational experience.

The observability lead role, owning the monitoring platform, evaluation frameworks, alert configuration, and incident review across all production models, becomes a full-time position at this maturity level. Organizations that have reached predictive intelligence have a named observability lead reporting to the VP of AI/ML, a monitoring runbook for each model class, and a quarterly review that connects observability data to retraining and retirement decisions. That operational structure is the hallmark of mission-aligned AI observability at portfolio scale.

Diagram 2.9 — AI Observability Maturity Progression

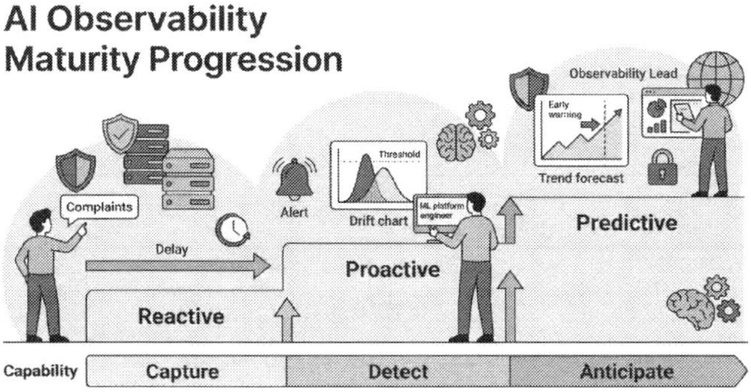

2.4 Manager's Checklist

- Confirm which production AI systems are covered only by APM tools and identify the model quality monitoring gap explicitly.
- Assign a named owner for model quality observability, distinct from the infrastructure SRE who manages uptime.
- Require that every production model's inference endpoint captures inputs and outputs to a queryable store before go-live.
- Document the current mean time to detect for the most recent model performance failure and use that figure in the observability ROI calculation.
- Define drift thresholds for each production model in collaboration with the model's business owner, not by statistical convention alone.
- Integrate AI observability alerts into the existing incident management workflow using PagerDuty or ServiceNow routing policies.
- Schedule a maturity self-assessment against the reactive-proactive-predictive spectrum for each production model.

- Require OpenTelemetry GenAI semantic convention compliance for any new AI system added to production.

2.5 Takeaway

Traditional DevOps observability is necessary but insufficient for AI systems. The three pillars, logs, metrics, and traces, provide complete infrastructure coverage and zero model quality coverage. Organizations that deploy AI in an APM-only monitoring environment are not instrumented to detect distribution shifts, accuracy degradation, or declines in output quality. AI observability extends beyond infrastructure health to the statistical and semantic properties of model behavior, and no amount of infrastructure tuning can close that gap.

The maturity spectrum from reactive to predictive defines the roadmap every organization must climb. Reactive detection is the common starting point; proactive monitoring is the minimum viable posture for any model managing consequential decisions; predictive intelligence is the target for organizations with growing portfolios. Each level requires specific tooling, processes, and accountability that must be planned and resourced deliberately.

3 Understanding What Can Go Wrong: Drift, Decay, and Degradation

A large retail bank deployed a loan origination scoring model, achieving 91 percent validation accuracy. By August of its first year, a quarterly review flagged a 12-point decline in accuracy. Investigation revealed three converging failure modes: an employment verification feature had drifted after a data partner changed its API schema; creditworthiness concept associations had degraded as pandemic-era payment patterns receded from the applicant population; and income quantile features had silently corrupted after a pipeline bug introduced rounding errors at a specific income range. No infrastructure alert had fired for any of these failures. Each was a distinct category of AI system degradation with its own mechanism, detection approach, and remediation path. The team had treated them as a single monitoring problem; they are three separate observability disciplines.

Every production AI system degrades over time. That is not pessimism; it is a structural property of statistical models operating in dynamic environments. The question is not whether a model will degrade but which failure mode will manifest first, how fast it will accumulate, and whether the organization will detect it before it reaches a consequential impact threshold. Managers who understand the taxonomy of AI failure modes can design monitoring programs that target each mode with the right detection method, rather than building a generic alerting strategy that covers none of them reliably.

The financial and operational exposure from undetected model degradation scales with portfolio size. An organization with 10 production models faces a complex risk landscape that requires a structured approach to monitoring prioritization. Models managing high-stakes decisions, including credit, fraud, clinical triage, and dynamic pricing, carry asymmetric risk: the cost of missing a degradation event is substantially higher than the cost of investigating a false positive alert. Building a monitoring program calibrated to that asymmetry requires understanding each failure mode's mechanism and business impact profile.

3.1 Data Drift and Feature Drift

Data drift and feature drift describe two closely related but distinct phenomena. Data drift refers to shifts in the statistical distribution of raw input data over time. Feature drift refers to shifts in engineered features derived from raw data, which may occur due to natural data shifts, pipeline changes, or upstream data source modifications. Both degrade model performance through the same mechanism: the model was optimized for a distribution it is no longer seeing.

When the Real World Moves but Your Model Does Not

Data drift occurs when the statistical distribution of features in production diverges from the training distribution. The training dataset is a historical sample collected under specific conditions. Production data reflects the world as it exists today, which may differ due to seasonal patterns, economic cycles, demographic shifts, or external events. When those differences accumulate beyond the model's tolerance, its learned mappings become miscalibrated, and prediction quality declines.

Operational detection of data drift relies on statistical distance measures. The Population Stability Index is widely used; a PSI above 0.2 conventionally signals significant drift. The Kolmogorov-Smirnov test compares cumulative distribution functions of training and live data and returns a p-value indicating whether they are drawn from the same distribution. Wasserstein distance measures how much probability mass must move to transform one distribution into another. Platforms like Arize AI, Evidently, and Monte Carlo implement these statistics on automated schedules and surface alerts when thresholds are crossed.

Managers should require that every production model have a documented feature-monitoring configuration specifying which features are monitored, which statistical test is applied, the alert threshold, and how often the check runs. The observability lead should be able to show a dashboard that shows the current drift status for each feature, a timestamp of the last evaluation, and a color-coded status indicator. Without that granularity, a drift alert provides no immediate guidance on which features are responsible, slowing diagnosis and extending the incident resolution window.

Diagram 3.1 — Data Drift Detection Pipeline

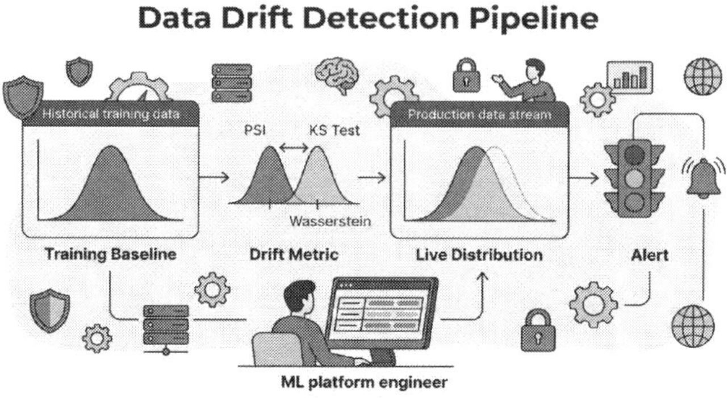

Feature Drift from Pipeline Corruption

Feature drift is particularly insidious when it originates not from natural shifts in the data distribution but from changes in the data pipeline that produce engineered features. An upstream data vendor changing its schema, a transformation function being updated in the feature store, a new data partitioning logic introduced by a data engineering team, or a silent bug in a data quality check can all alter the statistical properties of features that the model receives without any change in the underlying phenomenon the model is trying to measure. The model receives different numbers while believing it is receiving the same measurement.

Snowflake and Databricks environments, where enterprise feature engineering runs are complex, multi-team systems, are complex, multi-team systems where changes accumulate faster than change management processes can track. A data engineering team upgrading a dbt model or modifying an Apache Iceberg table schema may not realize the transformation affects a downstream feature that a production ML model depends on. Without end-to-end feature lineage instrumentation, the connection between a pipeline change and a performance decline is invisible. Tools like Monte Carlo and Acceldata detect schema changes and distribution anomalies in pipeline outputs before they reach the model serving layer.

The discipline that prevents pipeline-induced feature drift is a formal data contract between the feature engineering team and the model serving team that specifies expected schema, distribution bounds, freshness window, and notification requirements. When feature store output deviates from the contract specification, an automated quality check flags the deviation before the model consumes corrupted features. This contractual approach transforms feature drift detection from reactive forensics into proactive governance. Both teams sign off at deployment time, and both are accountable for any contract violations.

Diagram 3.2 — Feature Pipeline Data Contract Architecture

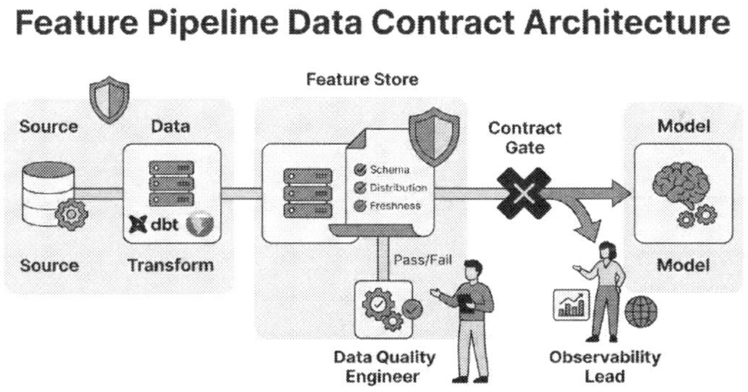

Detecting Multivariate and Interaction Drift

Most drift monitoring programs focus on univariate drift: tracking each feature's distribution independently and alerting when any single feature exceeds its threshold. Univariate monitoring is a necessary starting point, but it misses multivariate drift, where no individual feature distribution has changed significantly. Still, the joint distribution of features has shifted in ways that affect model performance. A credit model relying on the interaction between income and employment tenure may degrade without any univariate alert firing if both features drift modestly, but their correlation structure changes.

Detecting multivariate drift requires dimensionality reduction techniques that project the feature space into a lower-dimensional representation and track distribution shift there. Principal component analysis, combined with distribution testing on component scores, is an established approach. The Maximum Mean Discrepancy statistic, implemented in Evidently and the Alibi Detect library, provides a kernel-based multivariate shift test that does not require feature space reduction. For models where feature interactions are central to predictive power, multivariate drift detection is a required monitoring component.

Tree-based ensemble models, such as gradient-boosted trees, are particularly sensitive to interaction drift because their predictions depend on feature interactions captured in split conditions. Deep learning models that process feature embeddings can exhibit multivariate drift entirely invisible to per-feature tests. Models in high-stakes categories, specifically those driving credit, clinical, public safety, or revenue-sensitive decisions, should receive multivariate drift monitoring as a standard element of their production monitoring configuration, not an optional enhancement applied only when a univariate alert has already fired.

Diagram 3.3 — Univariate vs Multivariate Drift Detection

Univariate vs Multivariate Drift Detection

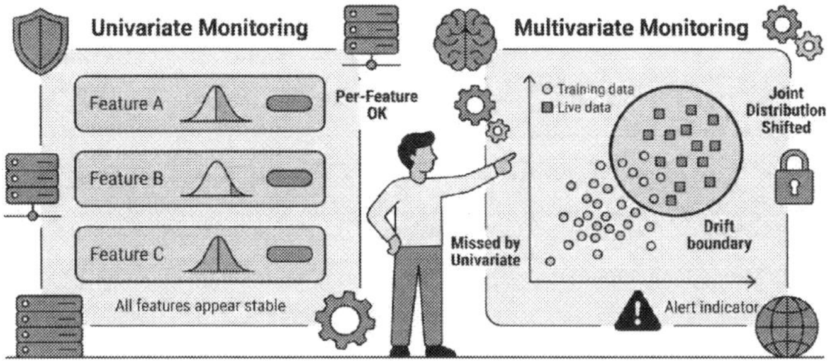

3.2 Concept Drift and Model Decay

Concept drift and model decay describe failure modes operating at a deeper level than input distribution shift. Concept drift occurs when the fundamental relationship between input features and the target outcome changes. Model decay is the broader phenomenon in which a model's predictive power declines over time. Both are inevitable in long-lived production systems; both require specific detection strategies and management responses.

Concept Drift: When the Relationship Changes

Concept drift occurs when the fundamental relationship between input features and the target outcome changes, invalidating the model's learned mapping regardless of whether feature distributions have shifted. A fraud detection model trained on pre-pandemic transaction patterns may find its discriminative features becoming less effective as spending patterns and fraud tactics both shift. The feature distributions look identical to the training baseline, and statistical drift tests show no alert, but model accuracy declines because the associations between features and outcomes have changed.

Two forms of concept drift require different detection strategies. Sudden concept drift occurs when a discontinuous event abruptly changes the relationship between features and outcomes; a regulatory change or major external event can cause it. Gradual concept drift occurs when the relationship shifts slowly, as in a churn model where factors that predict churn evolve incrementally as product features and competitive alternatives change. Sudden drift is more readily detected with threshold-based monitoring; gradual drift requires trend analysis across extended rolling windows.

The practical detection mechanism for concept drift requires outcome data: ground-truth labels confirming whether predictions were correct. When outcome data are available with a reasonable lag, monitoring programs can compute rolling accuracy metrics and alert when accuracy falls below defined thresholds. Tools like Fiddler AI and Arize AI support outcome-linked monitoring, comparing predictions against labeled ground truth as it arrives. When outcome data is delayed, as in many credit and insurance applications, prediction confidence distribution and output distribution stability serve as proxies for leading indicators.

Diagram 3.4 — Concept Drift Detection Framework

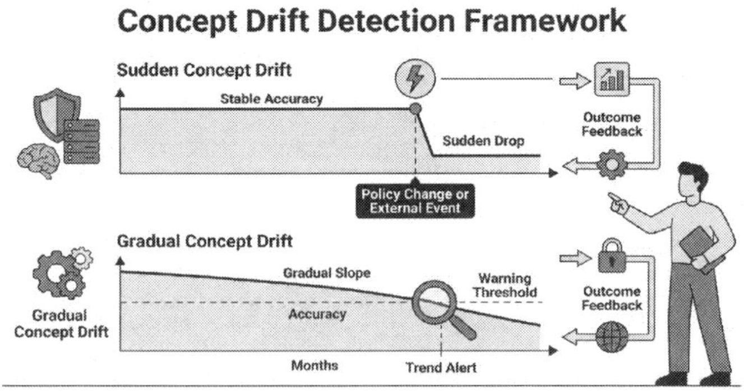

Model Decay: The Natural Half-Life of Production AI

Every production AI model has a half-life: a characteristic time horizon over which its predictive power degrades to the point that it no longer justifies deployment. That half-life varies by model class, domain, and environmental volatility. A real-time fraud detection model in a dynamic attack environment may have a half-life measured in weeks. A demand forecasting model for a stable product may remain calibrated for months. Knowing the expected half-life of each production model is a prerequisite for setting appropriate monitoring and retraining cadence.

The cumulative effect of multiple drift mechanisms drives model decay. Data drift narrows the model's applicable input range. Concept drift erodes the validity of its learned associations. Feature drift corrupts the measurements it depends on. Population shift, where the demographic or behavioral mix of the user population changes, alters the model's effective coverage. Each mechanism contributes independently to the overall decay trajectory, and in complex real-world deployments, they typically coexist. The monitoring program must track indicators of each to provide a comprehensive decay signal.

The management response to model decay is a retraining-and-retirement framework defined before deployment. The ML platform lead should document, for each production model, the expected decay trajectory, the monitoring metrics that serve as decay indicators, the threshold values that trigger retraining review, the retraining process, including data requirements and validation criteria, and the retirement criteria if retraining does not restore acceptable performance. This framework transforms the retraining decision from a reactive emergency into a governed, planned process with clear ownership and documented criteria.

Diagram 3.5 — Model Decay Lifecycle and Retraining Trigger

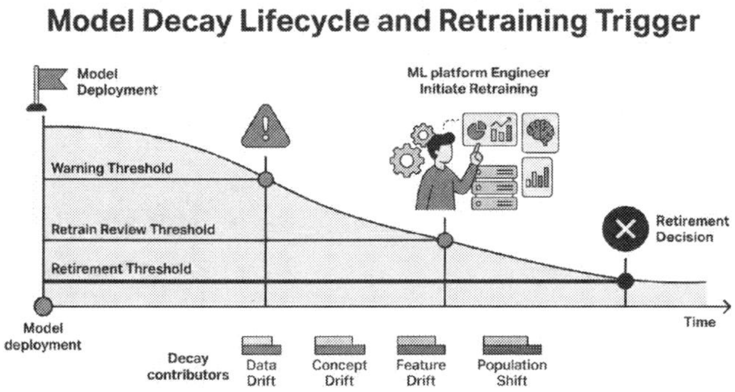

LLM-Specific Failures: Hallucination Drift and Prompt Sensitivity

Large language models introduce failure modes with no direct analog in traditional predictive ML. Hallucination drift describes the phenomenon in which a deployed LLM's rate of factually incorrect, plausible-sounding outputs increases over time or across shifts in input distribution. An LLM fine-tuned on an enterprise knowledge base may hallucinate at an acceptable rate during validation, but begin generating incorrect product information as user query patterns evolve toward topics underrepresented in the fine-tuning data. The rate of hallucination is not a fixed model property; it is a function of the query distribution, and it drifts as that distribution drifts.

Prompt sensitivity drift occurs when changes in prompt construction logic, or gradual shifts in how users phrase queries, cause the model to produce systematically different output distributions for semantically equivalent requests. An LLM serving a customer service application may handle a specific intent well when phrased one way and poorly when the same intent is expressed differently by a new user cohort. Monitoring prompt sensitivity requires a library of canonical test cases evaluated against the live model on a scheduled cadence, with alerts when output quality on canonical queries degrades.

Context window degradation is a third LLM-specific failure mode in retrieval-augmented generation deployments. As the document corpus grows, ages, or is updated, retrieved context quality can degrade in ways that cascade into degraded LLM output quality. Monitoring requires tracking retrieval quality metrics, including relevance scores and the faithfulness of outputs relative to the retrieved context, using evaluation frameworks such as Langfuse, LangSmith, and Ragas. Managers overseeing LLM deployments should require explicit monitoring configurations for hallucination rate, prompt sensitivity, and retrieval quality before production approval.

Diagram 3.6 — LLM Failure Mode Monitoring Coverage

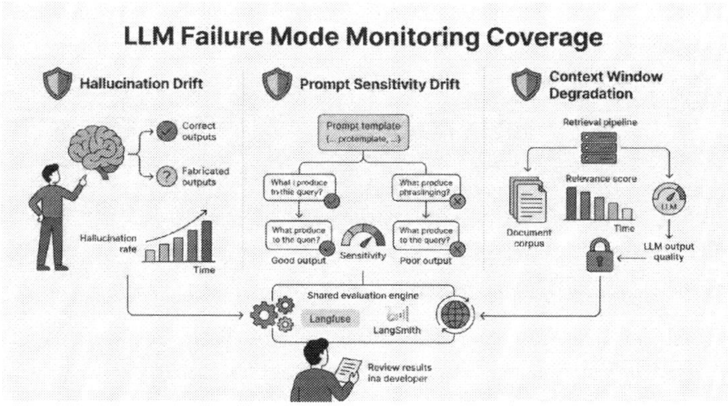

3.3 Business Impact Patterns and Monitoring Priority

Understanding failure mode taxonomy is necessary but not sufficient. Managers must connect each failure mode to its specific business impact pattern so that monitoring resources are allocated proportionally to business risk. Not all drift is equally consequential. A structured impact analysis performed at deployment time and reviewed quarterly enables observability, leading to prioritizing monitoring investment and calibrating response urgency for each failure mode type.

Revenue-Impacting vs Risk-Amplifying Failure Modes

Different failure modes result in distinct patterns of business exposure. Data drift in a recommendation engine typically manifests as a gradual decline in conversion rates, which business analytics can detect before drift monitoring fires an alert. For revenue-driving models, this lead-lag relationship argues for connecting business analytics dashboards to the model monitoring layer through a shared observability platform rather than treating them as separate systems. The business signal drives the technical signal, and both must be visible within a unified operational view.

Concept drift in risk models creates a different impact pattern: a quiet accumulation of misclassified decisions, surfacing as aggregate anomalies in loss or default rates. The business impact of concept drift in credit or fraud models is often asymmetric: the cost of false negatives, approving loans that should be declined, is typically far higher per incident than the cost of false positives. This asymmetry means concept drift in risk models warrants tighter thresholds and shorter detection windows than an equivalent accuracy decline in a marketing optimization model.

Feature drift due to pipeline corruption can lead to legally consequential outcomes regardless of the model's business category. A corrupted feature in a hiring AI tool may systematically exclude qualified candidates from a demographic group without intent. A corrupted feature in clinical decision support may alter risk stratification for a patient population. These are documented failure modes in deployed enterprise systems. The discipline of data contracts and pipeline observability tools like Monte Carlo is not merely an engineering best practice; it is a governance requirement for models in regulated or high-stakes domains.

Diagram 3.7 — Failure Mode Business Impact Profiles

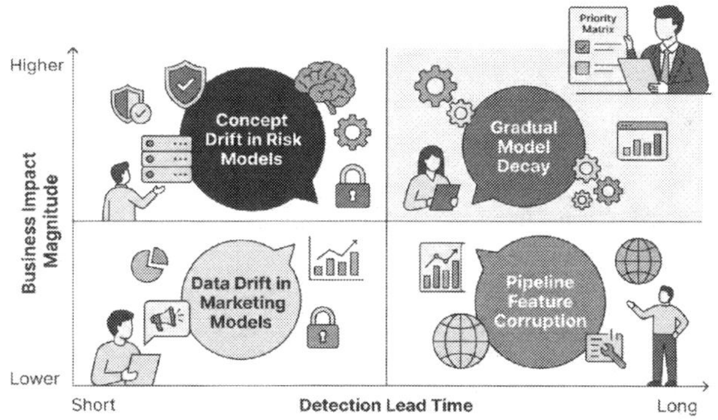

Setting Drift Thresholds That Align with Business Tolerances

Every drift threshold is simultaneously a statistical parameter and a business decision. Setting a PSI threshold at 0.1 rather than 0.2 doubles alert sensitivity but also increases the false-positive rate, requiring investigation. The correct threshold reflects the organization's tolerance for undetected degradation balanced against alert fatigue. A model driving autonomous credit decisions under regulatory reporting requirements should have tight thresholds and accept a higher false-positive rate for investigations. A model that informs non-binding analyst recommendations can tolerate wider thresholds without significant risk.

The threshold-setting process should involve three parties: the ML platform engineer, who understands statistical properties and can estimate false-positive rates across different settings; the business owner, who understands the cost asymmetry between missed detection and unnecessary investigation; and the compliance function, who understands regulatory requirements. This three-party consultation ensures thresholds are not set in isolation by engineers lacking business context or by business owners lacking statistical knowledge. The resulting configuration should be documented with the business rationale for each setting.

Review and revision of drift thresholds should occur during the quarterly model health review, based on the observed alert history. Thresholds that consistently fire false positives should be loosened with documented justification. Thresholds that fired too late in historical incident reviews should be tightened. This living configuration practice, treating thresholds as calibrated parameters that evolve with operational experience, is a hallmark of a mature observability program. The observability lead should maintain a change-log threshold for each production model as an auditable governance artifact.

Diagram 3.8 — Drift Threshold Calibration Process

Real-World Failure Examples and Their Business Cost

The abstract taxonomy becomes concrete through documented production failures. A major bank's mortgage approval model experienced gradual concept drift over eighteen months as post-pandemic economic normalization shifted the relationship between employment stability features and default rates. The model continued approving applications at validation-time accuracy until an audit revealed a 23-basis-point increase in early-payment defaults in the AI-assisted cohort. The investigation confirmed that the model had operated with stale concept associations for nine months before detection, requiring a restatement in the board's model risk report.

In enterprise SaaS, a customer success platform's churn prediction model experienced data and feature drift after a major product redesign altered the behavioral signals that had historically correlated with churn risk. Engagement features that had been strong predictors became noisy and weak after the redesign, but the feature transformation pipeline continued serving them as if their predictive value were unchanged. The model's predictions for at-risk accounts became miscalibrated, and the customer success team allocated renewal resources based on those predictions for two quarters before the drift was traced back to the redesign event.

These examples share a common pattern: each failure mode was detectable in principle at onset, but the organization lacked the monitoring infrastructure to detect it in practice. In each case, the business cost, including credit losses, customer churn, and management time spent on retrospective investigation, substantially exceeded the cost of the observability tooling that would have caught the drift within days. Managers should ask for each production AI system: if this failure mode were occurring today, how long would our current monitoring take to detect it?

Diagram 3.9 — Failure Mode Detection Gap Analysis

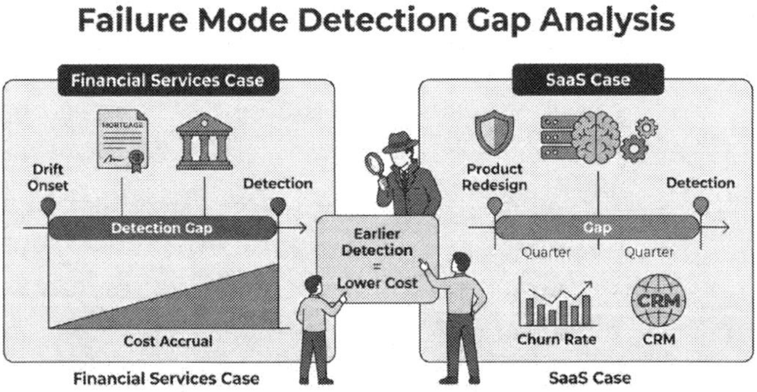

3.4 Manager's Checklist

- Require a documented failure mode analysis for every production model, identifying the primary risks: data drift, concept drift, feature drift, and model decay.

- Set univariate drift thresholds for each input feature in every production model, validated by both the ML platform engineer and the model's business owner.
- For models driven by feature interactions, require multivariate drift monitoring in addition to per-feature univariate monitoring.
- Establish a data contract between the feature engineering team and each model's serving layer, specifying schema, distribution bounds, and freshness requirements.
- Define the expected decay half-life for each production model and set a retraining review trigger at the appropriate degradation threshold.
- For LLM deployments, require monitoring configurations for hallucination rate, prompt sensitivity, and retrieval quality before production approval.
- Maintain a threshold change log for each production model so calibration decisions are auditable governance artifacts.
- Conduct a quarterly drift review that answers: if a significant failure mode were occurring in each production model today, how long would our monitoring take to detect it?

3.5 Takeaway

Drift, decay, and degradation constitute a taxonomy of distinct failure modes, each with its own mechanism, detection method, and business-impact profile. Data drift and feature drift track changes in what the model receives. Concept drift and model decay track changes in the model's knowledge. LLM-specific failures introduce semantic dimensions that statistical tests alone cannot measure. A monitoring program designed without this taxonomy will leave structural gaps.

The business case is a simple calculation: monitoring infrastructure costs are fixed, while undetected degradation costs scale with the detection gap. Every day a failure mode goes undetected in a high-stakes production model, it accumulates business exposure. Managing that exposure is the joint responsibility of the ML platform lead, the observability lead, and the VP of AI/ML.

4 The AI Observability Stack

A financial services firm's VP of AI/ML approved an AI-native monitoring platform after three silent drift events surfaced in a quarterly review. When the implementation team began integration, they found a fundamental architecture problem. The loan scoring model ran on a GPU cluster with only basic CloudWatch metrics. The fraud-detection LLM had token-level logging but lacked output-quality evaluation. The document processing pipeline included data quality checks that were not linked to model performance. The customer service agent ran in a vendor SaaS environment with non-exportable telemetry. There was no unified trace from data source to business outcome for any system. The team spent six weeks mapping architecture gaps before monitoring was possible. This is the experience of nearly every enterprise that attempts AI observability without first understanding the stack it is observing.

An AI observability program built without a clear architecture picture will fill gaps opportunistically rather than comprehensively, leading to integration surprises that delay deployment and inflate costs. The AI observability stack is not a single tool; it is a set of distinct layers, each generating different signal types and each requiring different instrumentation approaches. Managers who understand the stack architecture can set realistic implementation sequencing, allocate budget to each layer proportionally, and hold vendors accountable to specific capability requirements rather than aspirational feature lists.

The architecture decision at the stack level has long-term organizational consequences. A stack built on open standards creates a vendor-neutral foundation that survives platform changes and portfolio growth. A stack built on proprietary telemetry formats from a single vendor creates lock-in that becomes expensive and fragile as the AI portfolio expands. Senior managers who participate in architectural decisions rather than entirely delegating them to the platform engineering team protect the organization from technical debt that compounds with every new AI system added to the portfolio.

4.1 Infrastructure and Data Pipeline Layers

The bottom two layers of the AI observability stack address infrastructure health and data pipeline integrity. These layers leverage mature monitoring capabilities that enterprise organizations have already deployed for traditional software. The AI-specific requirement is to ensure they are properly configured to emit the signals higher layers need and that their telemetry is linked to AI workloads through shared trace identifiers.

Layer 1: Infrastructure Monitoring for AI Workloads

Infrastructure monitoring for AI workloads extends traditional compute and network observability to cover the resource types specific to AI execution environments. GPU utilization is the most distinctive addition: underutilized GPUs waste capital expense while saturated GPUs introduce latency spikes that do not appear as errors. Tools like Datadog, Dynatrace, New Relic, and Grafana provide GPU utilization monitoring for NVIDIA hardware through the DCGM exporter and the OpenTelemetry metrics pipeline.

Memory pressure is a second infrastructure concern specific to AI workloads. Large language models, with multi-billion-parameter weights, loading into GPU memory, create utilization patterns very different from those of traditional web services. A model approaching GPU memory capacity may begin swapping to CPU memory, introducing latency that is invisible as an error but degrades the inference SLA. Monitoring the full memory hierarchy, GPU VRAM, CPU RAM, and disk I/O for model weight loading, provides early warning signals to right-size infrastructure before SLA violations occur.

Inference endpoint latency monitoring requires AI-specific decomposition. The total latency of an AI inference request includes network transit time, queuing time at the inference server, model loading time if not memory-resident, inference compute time, and post-processing time. Each component has a different remediation path. A spike in inference compute time may indicate hardware degradation; a spike in queuing time may indicate insufficient server capacity. Decomposed latency tracing through OpenTelemetry spans with AI inference semantic conventions enables engineers to diagnose issues without guessing at the root cause.

Diagram 4.1 — AI Infrastructure Monitoring Layer

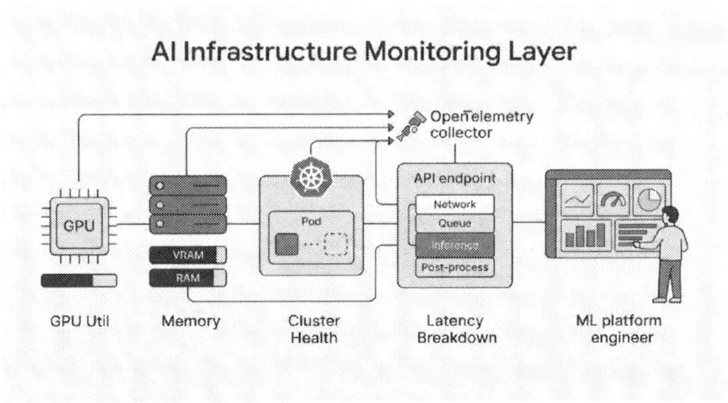

Layer 2: Data Pipeline Observability

The data pipeline layer collects, transforms, and serves input data to AI models. Failures in this layer are among the most common root causes of AI system performance issues, yet often the least well-observed. Data pipeline observability covers four dimensions: ingestion quality, schema integrity, data freshness, and feature lineage. Each requires dedicated monitoring that goes beyond what traditional ETL monitoring provides, because the AI model is sensitive to subtle statistical changes that row-count and job-completion checks miss.

Ingestion quality monitoring tracks whether incoming data's statistical properties match the expected distribution at each pipeline stage. A platform like Monte Carlo, Acceldata, or Great Expectations computes distribution statistics on each data artifact, compares them against a baseline, and alerts when deviations exceed tolerance. These checks run continuously upstream of the feature store and catch data quality issues before they reach the model. Ensuring data quality alerts route to the same observability dashboard as model performance metrics makes the connection between a pipeline anomaly and downstream model behavior immediately visible.

Feature lineage answers: for any given model prediction, which data sources, transformations, and feature engineering steps contributed to the input features? Without lineage instrumentation, root cause analysis for a feature drift event requires manual pipeline reconstruction, which is slow and error-prone. Platforms built on Apache Iceberg with integrated lineage tracking, or data catalog tools like Alation and Collibra with lineage graph capabilities, accelerate drift investigation from days to hours. The ML platform engineer should be able to trace any feature value to its source within the same observability session that surfaced the drift alert.

Diagram 4.2 — Data Pipeline Observability Dimensions

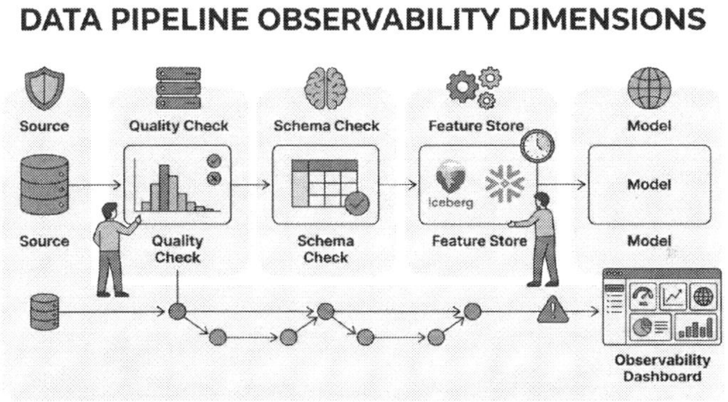

Connecting Infrastructure and Pipeline Signals

The value of the infrastructure and data pipeline layers is fully realized only when their signals are connected in a unified observability context. An infrastructure latency spike and a data pipeline freshness failure that occur simultaneously are almost certainly related; an observability platform presenting them in separate dashboards makes that connection difficult to see during a high-pressure incident. The architectural mechanism for unifying these signals is the distributed trace: a shared identifier accompanying a data record from ingestion through every transformation stage and inference event.

An OpenTelemetry-instrumented data pipeline propagates trace context from ingestion through feature engineering to the inference endpoint, creating a continuous span hierarchy that spans the entire data-to-prediction lifecycle. When an alert fires on model output distribution, the investigating engineer can navigate the trace to identify exactly where the relevant data deviation occurred, which pipeline job processed the affected records, and which infrastructure node was under abnormal load. An investigation that might take days using siloed monitoring tools takes hours using a unified OpenTelemetry trace.

Organizations that cannot immediately instrument the full pipeline with OpenTelemetry should establish correlation IDs at data ingestion and propagate them through the feature store to the inference log. Even without full trace instrumentation, a correlation ID linking a feature store record to a model prediction and then to the ground-truth outcome provides a rudimentary but functional lineage chain that enables post-incident root-cause analysis. ML platform leads should treat the absence of any correlation mechanism between pipeline data and model predictions as a monitoring gap that must be closed before a model reaches production.

Diagram 4.3 — Unified Trace Across Pipeline and Inference

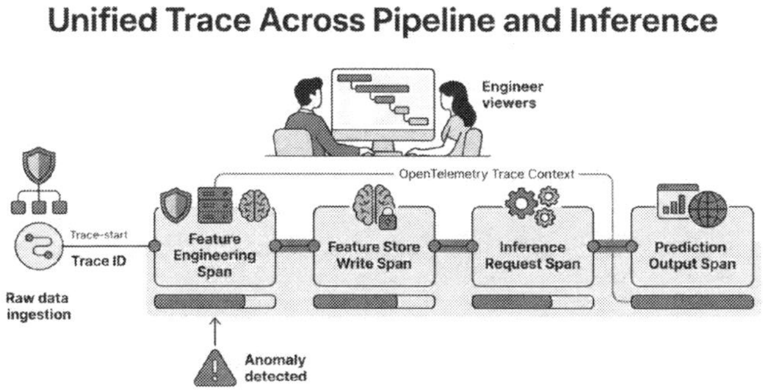

4.2 Model Performance and Output Evaluation Layers

Layers three and four of the AI observability stack address the questions that infrastructure and pipeline monitoring cannot answer: is the model still performing correctly, and are its outputs still aligned with the intended use case? These layers require AI-specific tooling and evaluation methods with no analog in traditional APM. They are also where the greatest observability investment gap exists in most enterprise organizations.

Layer 3: Model Performance Monitoring

Output evaluation is the layer that goes beyond statistical drift detection to assess whether model outputs are correct, safe, fair, and aligned with policy requirements. For traditional ML models, output evaluation is largely captured by performance metrics: if accuracy is within specification, outputs are within the intended quality range. For LLMs and generative AI systems, performance metrics are insufficient because correctness is a semantic property that statistics cannot capture. An LLM can produce fluent, grammatically correct, policy-compliant text that is nonetheless factually wrong; no statistical threshold catches that failure.

The LLM-as-a-judge approach uses a second AI model to evaluate the outputs of production models at scale. The evaluator is prompted with a rubric defining quality dimensions: factual accuracy, relevance, adherence to communication policy, absence of harmful content, and coherence across multi-turn conversations. The evaluator scores each sampled output on each dimension, and those scores aggregate into a quality dashboard monitored continuously. Tools like LangSmith, Langfuse, and Arize Phoenix provide evaluation pipeline infrastructure, including prompt management, sampling logic, and scoring aggregation required at production scale.

For traditional ML models, output evaluation is largely captured by performance metrics: if accuracy is within specification, outputs are by definition within the intended quality range. For LLMs and generative AI systems, performance metrics are insufficient because the output space is unbounded and correctness is a semantic property that statistics cannot capture. An LLM can produce fluent, grammatically correct, policy-compliant text that is nonetheless factually wrong; no statistical threshold catches that failure. Output evaluation exists to fill that semantic gap.

Diagram 4.4 — Model Performance Monitoring Framework

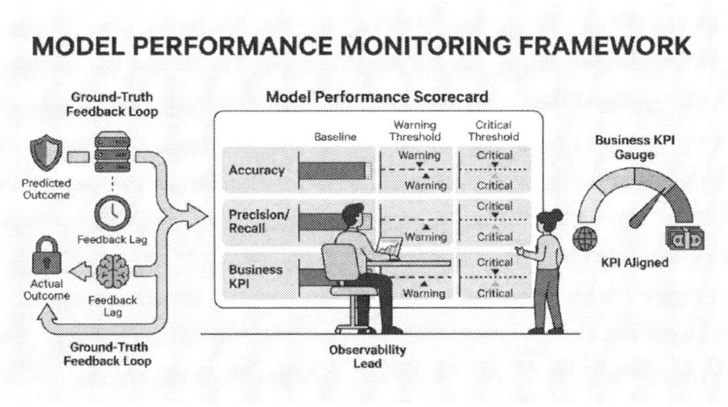

Layer 4: Output Evaluation for LLMs and Generative AI

Model performance monitoring tracks whether a model's predictive capability remains aligned with the validation baseline. Core metrics vary by model type: accuracy, precision, recall, and F1 for classification; mean absolute error and root mean squared error for regression; ranking metrics for recommendation models. Each metric must have a documented baseline value, a warning threshold below which investigation is triggered, and a critical threshold below which intervention is required. These three values form the monitoring specification for each metric and must be established at deployment time, not after the first degradation event.

The practical challenge of model performance monitoring is obtaining ground-truth labels in production. In fraud detection, a transaction can be confirmed or refuted within hours. In credit underwriting, the feedback loop may extend over 12 to 24 months. Monitoring programs in delayed-label environments rely on proxy metrics: prediction-confidence distribution and output-distribution stability, while ground-truth accuracy is computed on the subset of cases where labels are available. Hybrid monitoring designs combining proxy signals with lagged ground-truth evaluation are standard practice in financial services AI deployments.

Safety and toxicity monitoring is a distinct subset of output evaluation with a specific regulatory dimension. An LLM deployed in a customer-facing role is subject to reputational and regulatory risk if it produces harmful, biased, or policy-violating outputs. Guardrails monitoring tracks the rate at which production outputs trigger safety classifier alerts, the categories of safety violations that are occurring, and whether the rate is stable or trending upward. Platforms like Arize AI and Fiddler AI provide built-in safety evaluation modules that run alongside the primary quality evaluation pipeline. Organizations operating under HIPAA, financial services conduct regulations, or the EU AI Act requirements should treat guardrails monitoring as a compliance obligation rather than an optional quality enhancement, and the guardrails monitoring configuration should be documented in the model's governance record.

Diagram 4.5 — LLM Output Evaluation Pipeline

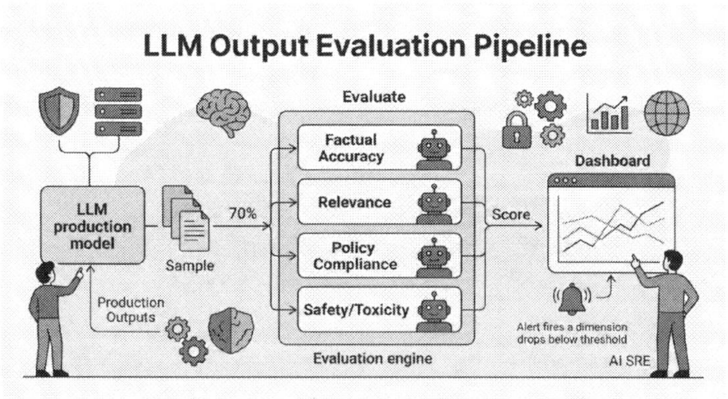

Layer 5: Agent and Workflow Observability

Agent observability addresses a requirement that did not exist in enterprise AI programs three years ago:

observing autonomous AI agents that make multi-step decisions, call external tools, and delegate subtasks to other agents. Agent observability extends trace-based monitoring to complex, branching execution graphs where a single user request may spawn dozens of tool calls, sub-agent invocations, and intermediate reasoning steps before producing a final output. Traditional span-based tracing is insufficient as the only observation mechanism for these non-linear, non-deterministic execution patterns.

Decision graph visualization is the primary observability primitive for agentic AI systems. A decision graph captures the full execution tree of an agent's response: the initial reasoning step, each tool call with its inputs and outputs, each sub-agent delegation, and the final synthesis step. This visualization enables the observability engineer to see exactly what the agent did at each decision point, which is the diagnostic information needed to investigate any agent failure. Tools like LangSmith, Langfuse, and Arize Phoenix provide decision graph capture as core features of their agent observability modules.

The tool called fidelity monitoring tracks whether an agent uses the right tools with the correct inputs. Monitoring invocation distributions and tool-call error rates across sessions, and alerting when patterns deviate from the evaluation baseline, provides the observability coverage that agent systems require beyond simple output-quality measurement.

Diagram 4.6 — Agentic AI Observability Components

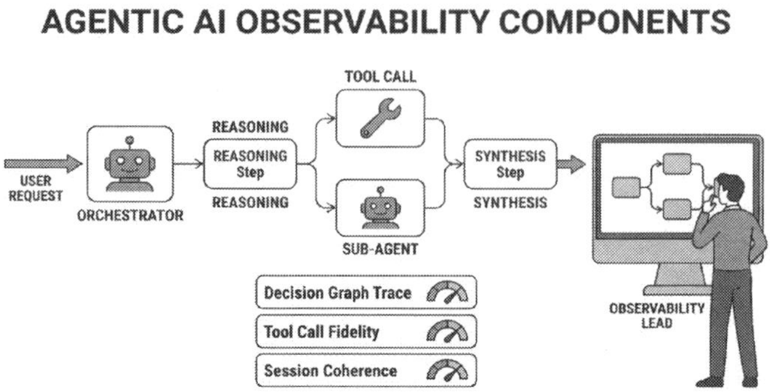

4.3 Open Standards and End-to-End Traceability

The five layers of the AI observability stack generate compounding value when connected by a unified data model and open telemetry standards. The connection is not automatic; it requires deliberate architectural decisions about telemetry collection, trace context propagation across layer boundaries, and observability data storage and querying. Managers who understand these principles can guide the platform engineering team toward implementation choices that build long-term capability.

OpenTelemetry as the Unifying Standard

OpenTelemetry is the CNCF project that defines a vendor-neutral standard for collecting, processing, and exporting telemetry, including logs, metrics, and traces. It is supported natively by AWS, GCP, Azure, Datadog, Dynatrace, New Relic, Grafana, and most AI-native observability platforms. For AI observability programs, OpenTelemetry provides a common telemetry language that connects infrastructure signals, data pipeline signals, model serving signals, and agent execution signals in a single trace hierarchy, without requiring each layer to use the same proprietary monitoring tool.

The OpenTelemetry GenAI semantic conventions extend the standard with AI-specific span attributes: model name, model version, input and output token counts, inference latency, and semantic evaluation scores. When every AI system in the portfolio emits telemetry using these conventions, the observability platform can provide cross-model analytics and portfolio-level health dashboards without per-model custom configuration. Requiring GenAI semantic convention compliance for every new AI system is an architectural governance decision that pays dividends at portfolio scale.

The Collector is particularly important for AI workloads because it enables enrichment of inference spans with additional context, sampling high-volume traffic, and fan-out of telemetry to multiple backends simultaneously. An ML platform engineer can route infrastructure metrics to Datadog, model performance spans to Arize AI, and compliance audit logs to a data lake from a single instrumentation effort. The Collector pattern is the architectural component that makes OpenTelemetry scalable to enterprise AI portfolio sizes without requiring per-application backend configuration.

Diagram 4.7 — OpenTelemetry AI Stack Integration

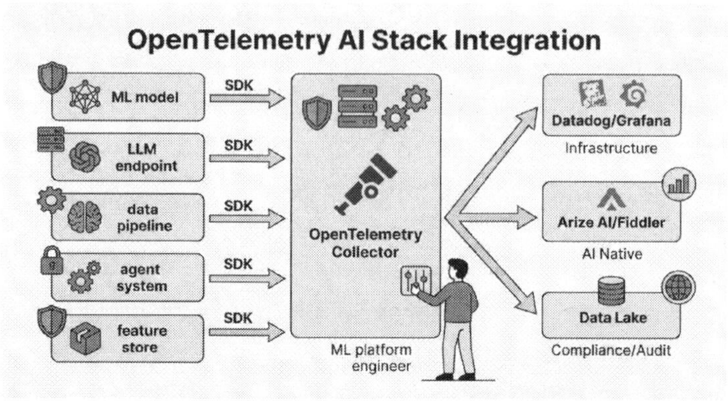

End-to-End Traceability from Data Source to Business Outcome

End-to-end traceability connects raw data at ingestion to the business outcome that results from the AI system's output. For any business outcome, an analyst can trace backward through the model prediction, the feature values the model received, the pipeline transformations, and the data quality checks to the source record that originated the prediction. This traceability chain is the audit record that regulators, model risk committees, and governance functions require to verify that AI system decisions are grounded in valid, documented data.

Constructing end-to-end traceability requires a consistent identifier strategy. Every data record entering the AI system must be assigned a unique identifier upon ingestion. That identifier must be propagated through every pipeline transformation, embedded in the feature store record, included in the inference request, and recorded alongside the model prediction. When the business outcome is observed, the same identifier links the outcome back to the original prediction and its provenance chain. For non-OpenTelemetry-instrumented stages, a correlation ID column in each data table provides minimal viable traceability until full instrumentation is achieved.

The governance value of end-to-end traceability exceeds its operational debugging value for models in regulated domains. When a regulator asks an organization to explain a specific credit decision, the answer must include the feature values the model received, the data sources from which those features were derived, the model version that produced the prediction, and the threshold that translated the prediction into the decision. All of those elements are available in a properly traced AI system. Building end-to-end traceability at initial model deployment is far less expensive than reconstructing it under regulatory pressure after the fact.

Diagram 4.8 — End-to-End AI Traceability Chain

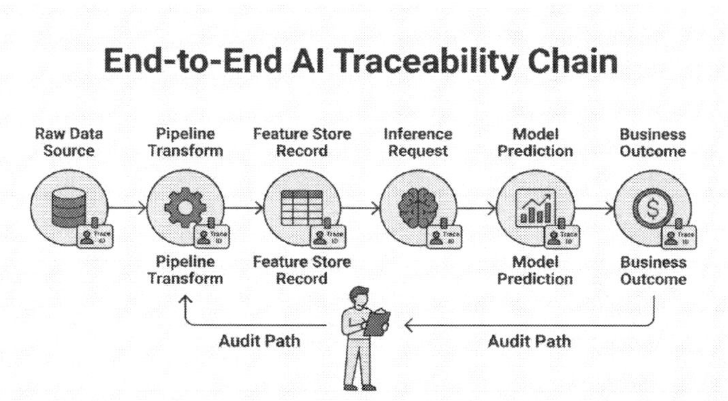

Reference Architecture for the Full AI Observability Stack

A production-ready AI observability architecture combines three tiers. The collection tier: OpenTelemetry SDKs in every AI application feeding a centrally managed Collector that performs sampling, enrichment, and routing. The storage tier: a time-series metrics backend like

Prometheus or Datadog, a trace store like Jaeger, and a data lake on Snowflake or Databricks for feature-level monitoring and ground-truth outcome records. The evaluation tier: an AI-native platform like Arize AI or Fiddler AI for model quality and drift monitoring, and an LLM observability tool like Langfuse or LangSmith for evaluating generative AI output.

The reference architecture is connected by the OpenTelemetry trace context, which propagates through every tier. A model prediction event generates a span carrying the model name, version, input features, and output. That span is stored in the storage tier alongside feature-level metrics and outcome feedback. The evaluation tier queries storage to compute drift statistics, evaluate output quality, and generate alerts. Alerts route to the on-call engineer through PagerDuty or ServiceNow, with a link to the specific trace, enabling navigation from the alert to feature values to the pipeline record in a single session.

Managers reviewing vendor proposals should evaluate each against this reference architecture. Key questions: Does the platform support OpenTelemetry ingestion natively? Does it provide trace-level navigation from alert to pipeline record? Does it integrate with the existing APM platform? Can it export data to the organization's data lake for compliance retention? Vendors who cannot answer these questions with specific integration patterns rather than roadmap aspirations should be deprioritized in the evaluation process.

Diagram 4.9 — Enterprise AI Observability Reference Architecture

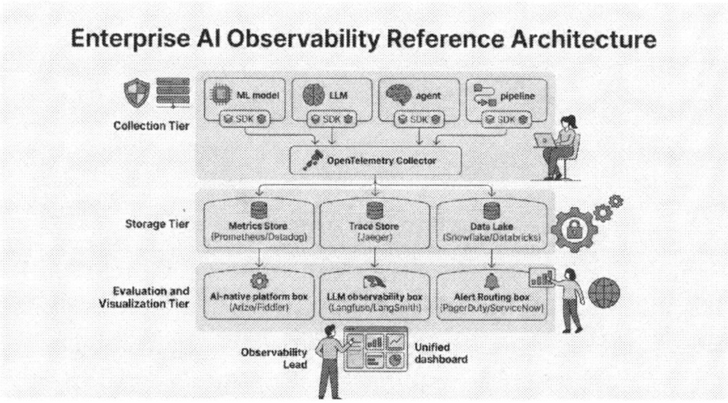

4.4 Manager's Checklist

- Require a stack architecture diagram showing all five observability layers for every new AI system before production approval.
- Mandate OpenTelemetry SDK instrumentation and GenAI semantic convention compliance for every AI application in the portfolio.
- Ensure data pipeline layer ingestion quality checks and feature lineage tracking are in place before model monitoring is deployed.
- Require that inference endpoint latency is decomposed into at least four components: queuing, loading, inference compute, and post-processing.
- For LLM deployments, confirm a continuous output evaluation pipeline using LLM-as-a-judge is operational before go-live approval.
- For agentic AI systems, require decision graph capture and tool call fidelity monitoring as non-negotiable observability components.
- Evaluate every AI observability vendor against the reference architecture using specific integration pattern questions, not roadmap features.

- Confirm that end-to-end trace IDs link raw data source records to model predictions and to business outcomes for every production AI system.

4.5 Takeaway

The AI observability stack is a five-layer architecture that delivers complete coverage only when all five layers are instrumented and connected. Infrastructure monitoring provides the foundation. Data pipeline observability catches upstream corruption before it reaches the model. Model performance monitoring tracks predictive capability against the validated baseline. Output evaluation assesses semantic quality that statistics alone cannot capture. Agent observability encompasses decision graphs and tool-call fidelity for autonomous systems.

OpenTelemetry and open standards unify the five layers into a navigable observability plane. Organizations that build on open standards protect themselves from vendor lock-in and create the audit trail that regulatory frameworks require. Building the stack correctly from the start is a management decision; retrofitting it under pressure is far more expensive in every dimension.

5 Monitoring Traditional ML Models in Production

A regional insurance carrier deployed a gradient-boosted model to score auto-loan applications in 2021. The model passed a formal promotion gate and was certified by the model risk team before going live. Eighteen months later, a compliance audit surfaced a troubling pattern: approval rates for a specific demographic segment had drifted by eleven percentage points relative to the training baseline. The ML platform engineer traced the shift to a feature-encoding change in an upstream data pipeline that had been propagating silently for six months. No alert had fired. The model was technically running; it was simply wrong. The carrier faced regulatory scrutiny, a remediation effort consuming three data science sprints, and a board-level briefing on model risk. The monitoring program the team thought they had was, in practice, a latency dashboard that confirmed responses without revealing whether decisions were correct.

Traditional predictive models, gradient-boosted trees, logistic regression classifiers, survival models, and neural networks built on structured features remain the workhorses of enterprise AI. They drive credit decisions, fraud alerts, churn predictions, demand forecasts, and clinical risk scores. When they degrade silently, the business impact is immediate and often invisible until it surfaces in financial results or a regulatory finding. Senior managers who treat model monitoring as a technical afterthought accept undisclosed risk on behalf of their organizations.

The stakes are concrete. A model that drifted by 15% on a key accuracy metric may still generate inferences at normal latency; the infrastructure dashboard shows green while the business absorbs losses. Regulators under SR 11-7 and the EU AI Act explicitly require ongoing performance validation for high-risk models. Failure to maintain documented monitoring programs can trigger mandatory remediation orders, civil penalties, and suspension of a model's use-case authority. Monitoring is not a best practice; it is the minimum condition for operating AI responsibly.

5.1 Establishing Baselines and Choosing the Right Metrics

Before any alert can fire, an ML platform team must answer a deceptively simple question: what does good look like for this model? The answer requires a formal baseline, a documented snapshot of input distributions, output distributions, and performance metrics captured at promotion time. Without a baseline, every monitoring signal is noise. With one, drift becomes measurable, promotions become defensible, and rollbacks become objective decisions rather than arguments.

The Baseline Distribution Report as a Promotion Artifact

A baseline distribution report is a structured document generated at promotion time that captures the statistical profile of every feature the model consumes: mean, standard deviation, percentile distribution, proportion of nulls, and distributional shape for continuous features; category frequencies and out-of-vocabulary rates for categorical features. This report becomes the reference against which all future production data is compared. Platforms such as Arize AI and Evidently AI generate these reports automatically when integrated with a model registry, but the artifact must be version-controlled alongside the model binary.

The practical value emerges during incidents. When a fraud detection model begins approving transactions that it would have rejected twelve months earlier, the first diagnostic question is whether the input feature space has shifted. An engineer who can pull the baseline report and overlay it on last week's production distributions can develop a root-cause hypothesis within minutes. Without a baseline, reconstruction from training data routinely consumes days. The baseline report standardizes forensic archaeology for comparison.

Managers should require that every model entering production carries a baseline distribution report as a non-negotiable promotion artifact. The model registry entry, whether managed in MLflow, SageMaker Model Registry, or Databricks, must include a pointer to the versioned baseline. Any promotion gate that does not verify its presence and completeness is a gate in name only. The data quality engineer and the ML platform engineer jointly own this artifact; neither can sign off on promotion until they confirm it is up to date.

Diagram 5.1 — Baseline Report at Promotion Gate

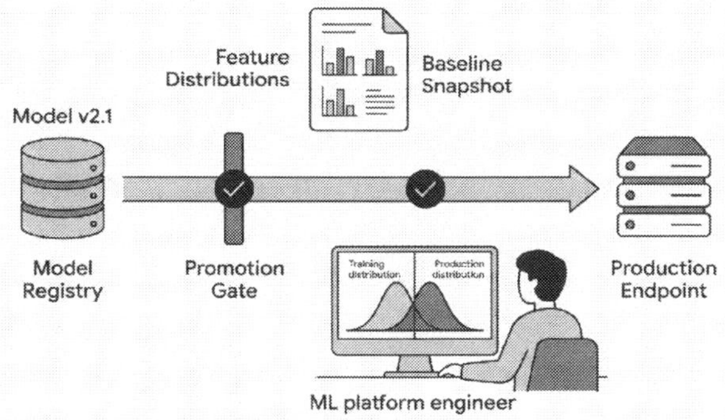

Selecting Metrics That Reflect Business Reality

Accuracy is a seductive metric because it is simple and universally understood, and it is also frequently the wrong one for a production monitoring program. A fraud classifier operating in a setting where 0.3 percent of transactions are fraudulent can achieve 99.7 percent accuracy by predicting every transaction as legitimate. A churn model evaluated solely by overall accuracy masks severe degradation in the high-value segment that the business cares about most. Monitoring metrics must reflect the actual business decision the model supports, not the metric that was easiest to optimize during training.

Precision and recall at the operating threshold, calibration error, and business KPIs such as false-positive cost and expected revenue impact belong on a production monitoring dashboard alongside raw accuracy metrics. For imbalanced-class models, the area under the precision-recall curve is more informative than ROC-AUC. For regression models, mean absolute percentage error by segment is more actionable than aggregate mean squared error. Fiddler AI allows operators to define custom business metrics that combine model output scores with downstream outcome data. Hence, a dashboard shows not only model performance but also its realized business impact.

The observability lead should convene a joint session with the model owner and a business stakeholder to define the primary metric, secondary metrics, and the business KPI translation layer at promotion time. This session produces a monitoring specification document listing each metric, its threshold, its alert tier, and the business interpretation of a threshold breach. This document accompanies the model throughout its lifecycle and is updated at each retraining event. Without it, monitoring dashboards become displays of numbers rather than instruments of operational clarity.

Diagram 5.2 — Metric Selection Decision Framework

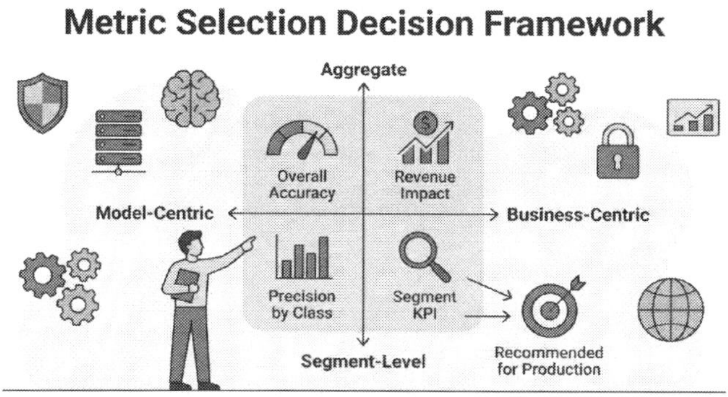

Real-Time vs. Batch Monitoring Architectures

Not all models benefit from the same monitoring cadence. A real-time fraud detection model scoring transactions in under 50 milliseconds requires streaming observability infrastructure computing feature statistics on rolling windows measured in minutes. A nightly batch job can monitor a demand-forecasting model that generates weekly purchase recommendations. Matching monitoring cadence to operational tempo is an architectural decision; over-engineering a batch model with streaming infrastructure wastes budget, while under-engineering a real-time model with nightly batch monitoring leaves hours of degraded decisions undetected.

Streaming monitoring pipelines for real-time models typically use a message bus such as Apache Kafka to capture inference requests and responses, feed them through a feature statistics computation layer, and write summary statistics to a time-series store that Grafana or Datadog can visualize. Arize AI and New Relic offer connectors that ingest streaming inference data with low overhead. The key engineering tradeoff is between window length and detection latency: a one-hour rolling window detects gradual drift, but a sudden spike in null rates caused by an upstream API failure requires a five-minute window.

Batch monitoring architectures are simpler but require careful comparison logic. Each nightly run must compare the current batch's feature and output distributions against the stored baseline, compute drift statistics, evaluate configured business metrics, and write results to the observability platform before the next cycle begins. Monte Carlo Data and Evidently AI both support scheduled batch profile comparison with automatic anomaly flagging. The AI SRE assigned to the model should review batch monitoring output as the first task each morning, most effectively when the platform surfaces a single summary health score alongside detailed statistics.

Diagram 5.3 — Real-Time vs. Batch Monitoring Paths

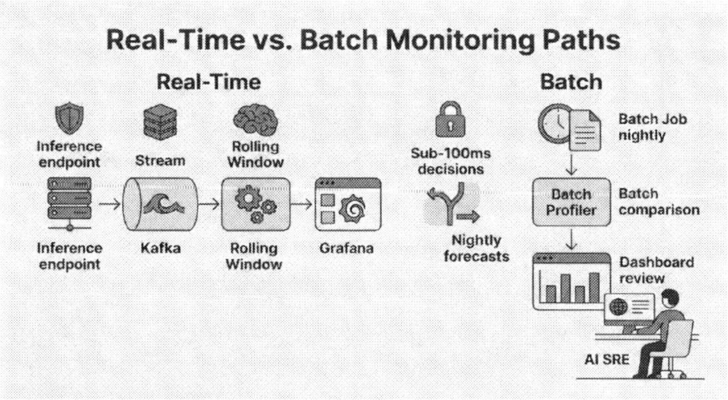

5.2 Drift Detection Methods and Alert Policies

Drift detection is the technical core of ML model monitoring. The question it answers is whether the statistical properties of the data flowing through a production model have changed enough to warrant concern. The decisions it drives are managerial: investigate, escalate, or retrain? Choosing the wrong statistical test, setting thresholds without a cost model, or building alert policies without tiering can result in either a program that fires constantly or one that never fires until a business crisis is already underway.

Statistical Tests for Feature and Prediction Drift

Three statistical methods dominate production ML drift detection. The Kolmogorov-Smirnov test compares empirical cumulative distribution functions of a reference window and a production window, returning a p-value indicating whether the two samples could plausibly share the same underlying distribution. The KS test is sensitive, distribution-free, and widely supported across monitoring platforms, including Evidently AI and Fiddler AI. Its limitation is that it weights all distribution regions equally; a tail shift in a credit score distribution may be more business-critical than a center shift, yet the KS statistic treats both identically.

The Population Stability Index quantifies distribution shift using an information-theoretic formula. A PSI below 0.1 indicates minimal shift; values between 0.1 and 0.2 indicate moderate shift requiring investigation; values above 0.2 indicate significant shift warranting retraining or escalation. PSI is popular in financial services, where model risk teams have decades of precedent for interpreting it. Wasserstein distance avoids PSI's dependence on binning by computing the minimum transport cost between two distributions. It is more computationally intensive but provides a more nuanced measure of distributional divergence, increasingly favored in modern monitoring frameworks.

An effective monitoring program applies KS tests to categorical features, PSI to continuous features where financial-services precedent matters, and Wasserstein distance to high-cardinality numerical features where tail sensitivity is critical. Each metric feeds into a separate alert channel with its own threshold, calibrated against the cost of the model's decisions. A model approving million-dollar commercial loans warrants tighter drift thresholds than a model generating e-commerce product recommendations.

Diagram 5.4 — Three Drift Statistics Compared

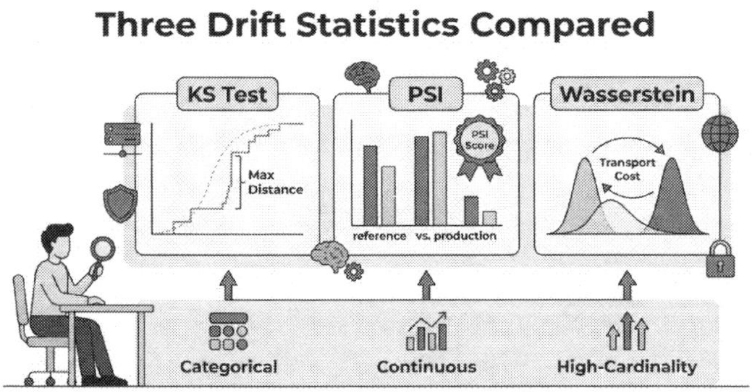

Setting Thresholds Without Creating Alert Fatigue

Alert fatigue is the graveyard of monitoring programs. When every drift statistic above a fixed generic threshold fires a page, the on-call AI SRE quickly learns that most alerts are false alarms triggered by seasonal variation, upstream schema changes, or batch timing artifacts. Within weeks, the team begins ignoring alerts; within months, the monitoring program exists only on paper. The antidote is a threshold-setting methodology that distinguishes statistical significance from operational significance and builds tiered alert policies rather than a single binary trigger.

Tiered alert policies assign each metric to one of three severity levels. Informational alerts fire when drift statistics exceed a low threshold and are logged to the observability dashboard without generating notifications; they serve as data points for the weekly model health review. Warning alerts fire at a higher threshold and send a message to the ML platform team's monitoring channel, requiring acknowledgment and a written hypothesis within 24 hours. Critical alerts fire when drift is severe or when a performance metric crosses a business-impact threshold, triggering an incident record that requires a response within four hours.

Thresholds should be calibrated through a three-month burn-in period during which the monitoring program runs in shadow mode, logging what would have fired without generating notifications. The burn-in data reveal the natural variability of each metric under normal operating conditions, allowing observability to inform setting thresholds at multiples of that variability rather than at arbitrary fixed values. Datadog and New Relic support anomaly-based alerting that automatically learns baseline variability. Every threshold must be documented, reviewed quarterly, and updated after any retraining event.

Diagram 5.5 — Tiered Alert Policy Structure

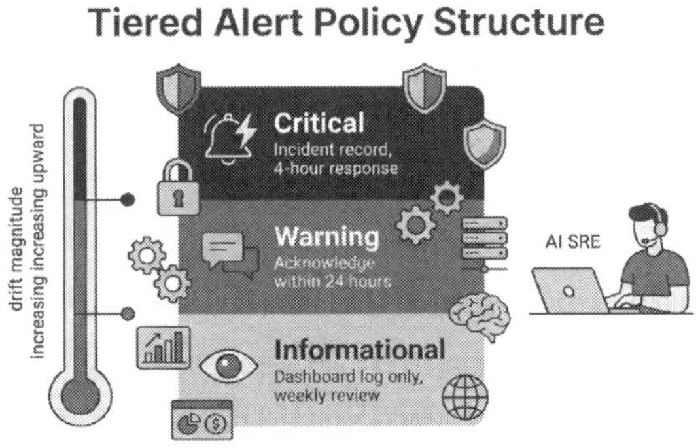

Monitoring Prediction Distributions and Output Confidence

Feature drift is the most commonly monitored signal, but prediction distribution drift is often the more immediately actionable one. When the distribution of raw model output scores shifts relative to baseline, it indicates the model's internal representation of the problem has changed, regardless of whether any individual input feature has drifted. A fraud model whose average score shifts from 0.12 to 0.08 over four weeks is almost certainly operating in a different risk environment, even if no single feature's PSI has crossed a warning threshold.

Monitoring prediction distributions requires the same statistical infrastructure as monitoring input features, but the business interpretation differs. A downward shift in average fraud probability during a period of known elevated fraud losses suggests the model is systematically underestimating risk. A gradual upward shift in average churn probability, without increases in the actual churn rate, may indicate the model is over-sensitive, inflating intervention campaign costs. Both patterns are invisible to infrastructure monitoring tools and require purpose-built

ML observability platforms, such as Arize AI or Evidently AI, to surface them.

Confidence calibration monitoring tracks the relationship between a model's stated confidence and its actual accuracy across bins. A well-calibrated model claiming 80 percent confidence in a decision class should be correct 80 percent of the time on that class. Calibration error, measured by expected calibration error or a reliability diagram, degrades over time as the model encounters inputs differing from the training distribution. Managers should require calibration monitoring for every model in which confidence scores are surfaced to downstream decision makers, because a decision maker trusting a miscalibrated score makes decisions based on false information.

Diagram 5.6 — Prediction Score Distribution Shift

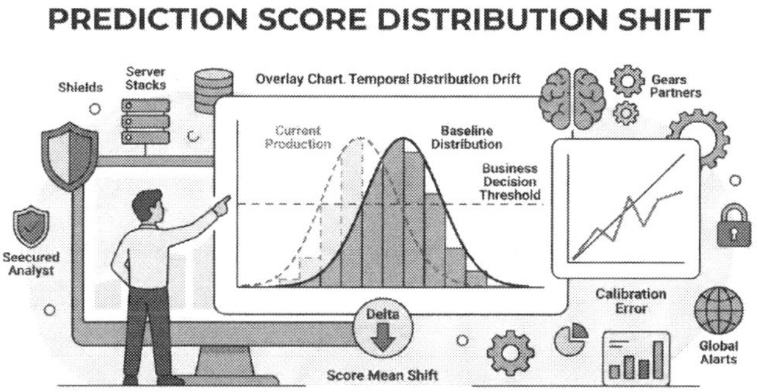

5.3 The Retraining Decision Framework and Model Lifecycle Governance

Detecting drift is necessary but not sufficient. The monitoring program must connect detected drift to a decision: investigate further, retrain, adjust the operating threshold, roll back to a previous version, or retire the model entirely. Organizations that have not pre-specified this decision logic spend valuable time debating what to do after an alert fires, often while the model continues to degrade. The retraining decision framework turns that debate into a documented workflow with clear ownership, defined criteria, and governed execution paths.

Defining Promotion Gates Across the Model Lifecycle

A promotion gate is a formal decision point at which a model version must satisfy quantified criteria before advancing to the next stage of the deployment lifecycle. Most enterprises distinguish four stages: training and offline evaluation, validation in a held-out environment, canary deployment on a fraction of production traffic, and full production rollout. Each transition is a gate with its own criteria. The gate from training to validation typically requires the model's offline metrics to exceed the baseline on a holdout set. The gate from canary to full production requires that live performance metrics meet the monitoring specification established at promotion time.

Promotion gates prevent two common failure modes. The first is premature promotion: a model pushed to production before its behavior under real-world traffic has been assessed due to a sprint deadline. The second is indefinite canary limbo: a model sitting in canary status for months because no one has specified what criteria would justify full promotion or rejection. Both are symptoms of missing governance. A well-governed promotion gate requires a formal approval record, signed by the model owner and reviewed by the model risk team.

Canary monitoring deserves particular attention. During canary deployment, the monitoring platform should run the new model version and the incumbent in parallel on split traffic, comparing prediction distributions, business metric outcomes, and latency profiles. Arize AI and Fiddler AI provide champion-challenger comparison capabilities for traditional ML models. The AI SRE responsible for the canary should generate a written comparison report at the end of each week of canary operation, documenting head-to-head metrics and recommending promotion or rollback.

Diagram 5.7 — Model Lifecycle Promotion Gates

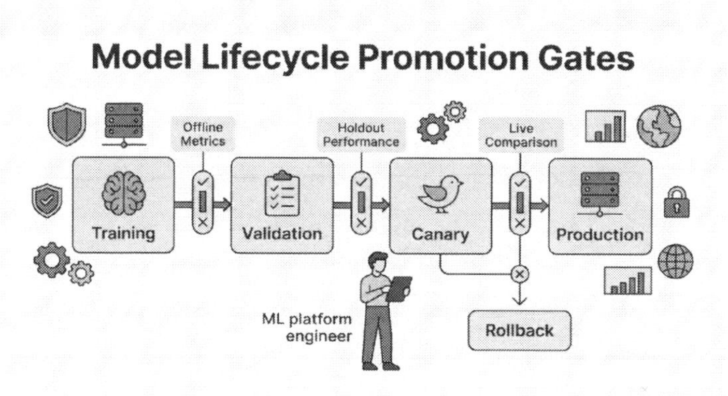

When to Retrain, Retune, or Retire

The retraining decision is among the most consequential and frequently mishandled decisions in the ML model lifecycle. Teams that retrain too aggressively consume expensive compute resources, introduce unnecessary variance, and create audit complexity as model versions multiply. Teams that retrain too infrequently allow degradation to accumulate until it becomes a business crisis. The correct trigger is a conjunction: a confirmed drift signal above the critical threshold, evidence that the drift is sustained, and a business impact assessment quantifying the cost of continued degraded performance against the cost of a retraining cycle.

Retuning, adjusting model hyperparameters, operating thresholds, or feature weights without a full retraining cycle, is appropriate when the model's learned representations remain valid, but calibration has shifted. A credit model drifted slightly in confidence calibration but maintained stable feature importance rankings, making it a candidate for threshold adjustment rather than full retraining. The data quality engineer should evaluate whether drift is driven by feature distribution shift or calibration shift before committing to the more expensive option. Threshold adjustment can often be executed within a single sprint; full retraining typically requires four to eight weeks.

Retirement is the least exercised option in most organizations and the most important one to normalize. A model whose use case has been superseded, or whose risk profile has become unacceptable under current regulatory standards, should be formally retired rather than left running as a zombie inference endpoint. Retirement requires a decommissioning record, notification to downstream consumers, and archival of all monitoring artifacts for the applicable retention period. The VP of AI/ML should review the active model inventory quarterly to identify candidates for retirement.

Diagram 5.8 — Retrain, Retune, or Retire Decision Tree

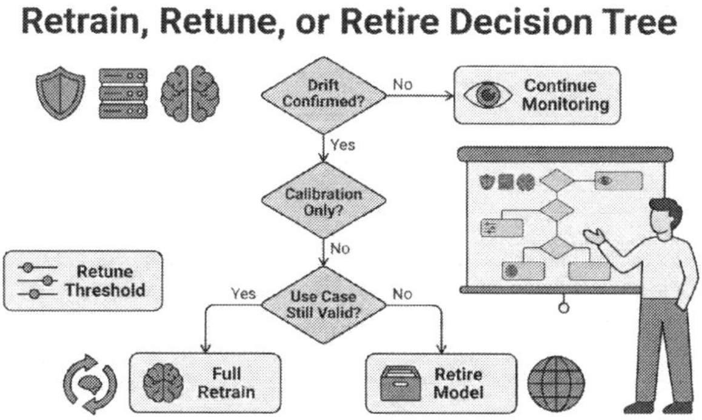

Maintaining Observability Continuity Across Environments

Model monitoring does not begin when a model reaches production. It begins the moment training completes, because the comparison between training and production behavior is the diagnostic foundation for every drift investigation. Organizations that maintain observability continuity across training, validation, canary, and production stages detect environmental discrepancies before they become production incidents. A model that performs measurably differently in validation than in training has a data leakage problem or a preprocessing inconsistency; catching it at validation is much cheaper than catching it six months into production.

Practical implementation requires a shared observability platform that ingests inference data from all deployment stages into a unified data store, tagged by environment and model version. OpenTelemetry semantic conventions, extended through the OpenTelemetry GenAI working group, provide a vendor-neutral instrumentation standard for tagging inference spans with environment, version, and feature metadata. Teams that instrument their inference code with OpenTelemetry from the start can route the same telemetry to a local dashboard, a validation monitoring environment, and a production platform such as Grafana or Datadog by changing only the export endpoint configuration.

Observability continuity also requires the baseline artifact to be propagated correctly through the lifecycle. When a model is retrained, the new baseline must be generated and version-controlled before the retrained model is promoted to validation. The canary comparison must use the incumbent model's production baseline, not the challenger's training baseline, because the comparison the business cares about is the challenger's behavior relative to the incumbent's current production. Misaligned baseline references are a common source of false confidence in canary comparisons.

Diagram 5.9 — Cross-Environment Observability Continuity

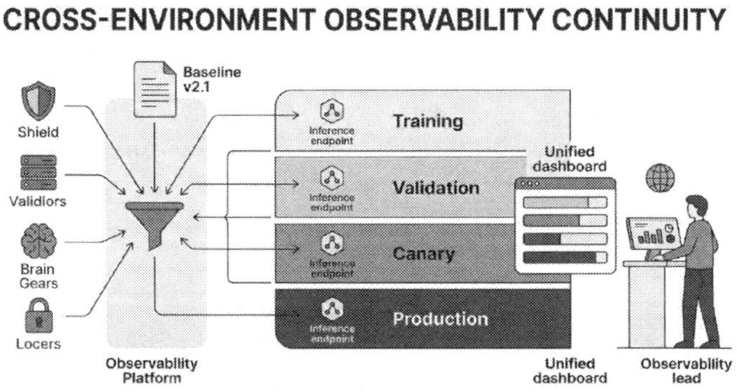

5.4 Manager's Checklist

- Require a baseline distribution report for every model before production promotion, stored in the model registry.

- Produce a monitoring specification document at promotion time listing each metric, its threshold, alert tier, and business interpretation.
- Match monitoring cadence to operational tempo: streaming for real-time inference models, nightly batch for forecasting models.
- Implement a three-tier alert policy (informational, warning, critical) with documented response times, calibrated during a 90-day burn-in period.
- Apply drift statistics appropriate to feature type: KS test for categorical, PSI for continuous, Wasserstein distance for high-cardinality numerics.
- Monitor prediction score distributions and calibration error for every model where scores drive material decisions.
- Require a written canary comparison report before full-production promotion, signed by the AI SRE and reviewed by model risk.
- Conduct a quarterly model inventory review to identify retirement candidates and formally decommission zombie inference endpoints.

5.5 Takeaway

Traditional ML models are not set-and-forget systems. Every model driving a material business decision operates in a changing environment, and without a monitoring program built on baselines, drift statistics, tiered alerts, and governed promotion gates, degradation is invisible until it becomes a crisis. Organizations that avoid those crises treat monitoring as a first-class engineering deliverable.

The practices here form the operational foundation for the more complex monitoring challenges ahead. LLM observability, agentic system monitoring, and vendor platform selection all build on the same discipline: define what good looks like, continuously measure deviation from it, and make the response to deviation a documented, governed decision.

6 Monitoring LLMs and Generative AI in Production

A financial services firm deployed a GPT-4-class language model in 2023 to summarize regulatory filings for its compliance analysts. Within three weeks, it had become the team's default tool for first-pass document review. Six months later, an internal audit found that summaries consistently omitted specific risk disclosure language. This behavior had shifted gradually as individual analysts updated prompt templates without version control. No token budget alarm had fired. No output quality threshold had been configured. The model was responding at normal latency and consuming tokens within the historical cost band. By every infrastructure metric, it was working. By every compliance metric, it was failing. That gap is the entire observability problem unique to large language models.

Large language models introduce a category of production risk that has no analog in traditional predictive ML. A gradient-boosted model either predicts a class or it does not; the failure mode is measurable. An LLM can generate plausible, confident, well-formatted output that is factually wrong, regulatory non-compliant, or subtly harmful, with no signal visible to infrastructure monitoring. The VP of AI/ML who treats LLM monitoring as an extension of existing model observability practices is operating with a significant blind spot.

The management stakes extend across five dimensions simultaneously: latency, cost, quality, safety, and compliance. A single production LLM application can violate all five without triggering an alert from a conventional APM platform. Token cost overruns can eliminate the business case for a use case overnight. A prompt injection session can expose customer PII. Output quality drift can erode analyst trust over weeks before it becomes a documented problem. These risks require a purpose-built observability approach, not a retrofit of traditional monitoring tooling.

6.1 Prompt Versioning, Token Economics, and the Five Dimensions of LLM Observability

Traditional ML observability collapses when applied to large language models because the inputs and outputs are natural-language artifacts rather than numerical vectors, and the relationship between them is governed by a prompt template whose evolution is typically uncontrolled. Before any quality signal can be measured or any cost attributed, the organization must solve the foundational problem of knowing precisely what instructions were given to the model at every inference call. That requires prompt versioning, which most enterprises lack.

Prompt Versioning as Observability Infrastructure

A prompt template is the primary control surface of an LLM application. It defines the model's persona, task instructions, output format constraints, safety guardrails, and the strategy for using the context window. When a prompt template changes, the model's behavior changes, often in ways that are non-obvious from inspecting the template diff alone. An analyst who adds a single sentence to a system prompt to improve formatting may inadvertently suppress a category of output that a downstream compliance check relies on. Without a prompt version control system, this change is invisible to the monitoring program.

Prompt versioning requires treating prompt templates as first-class code artifacts, stored in a version control system, tagged with semantic version numbers, and associated with the inference logs generated during each version's active period. Platforms such as Langfuse and LangSmith provide native prompt registry capabilities that capture the full system prompt, user prompt template, and model configuration parameters for each deployed version, and associate each inference call with the prompt version that generated it. This association is the prerequisite for every downstream observability signal.

The observability lead should establish a prompt change management process with the same rigor as a software deployment process. Prompt changes should require a pull request, a review from the application owner and a safety reviewer, and a canary deployment period during which the new prompt's output quality and safety metrics are evaluated against the previous version. This process is the control mechanism that makes LLM applications observable and governable. Without it, the prompt history of a production LLM application is a series of anecdotes rather than an audit trail.

Diagram 6.1 — Prompt Version Registry and Inference Trace

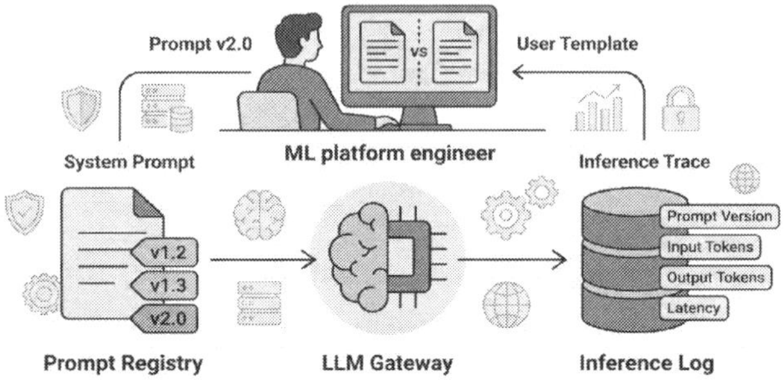

Token Budget Management and Cost Attribution

Token economics is the FinOps layer of LLM observability. A production LLM application can generate cost overruns of hundreds of thousands of dollars per month if token consumption is not instrumented, attributed, and governed. Cost visibility requires tracking input and output tokens separately at the level of individual inference calls, users, application features, and model versions. Organizations that track only aggregate monthly API spend have cost visibility but not cost observability.

Cost attribution at the feature and user level requires that each inference call carry metadata tags identifying the originating user, the feature or workflow that triggered the call, the model endpoint, and the active prompt version. Langfuse and LangSmith support structured metadata on inference calls that enable this attribution. The resulting data allows an observability lead to produce a weekly report showing token spend by feature, identifying the most expensive features, whether that cost is justified by usage volume, and whether any feature has experienced a token usage spike indicating a prompt efficiency regression.

Token budgets, pre-set limits on the tokens an application feature or user session may consume, are the operational mechanism for cost governance at the application layer. An enterprise deploying an LLM-powered customer service agent should set a per-session token budget reflecting the expected cost per interaction and the business value of that interaction. When a session's budget is approaching, the application should trigger fallback behavior: shortening the context, routing to a smaller, cheaper model, or gracefully ending the session. Datadog and New Relic support custom metric alerting that fires when token consumption rates cross pre-configured thresholds.

Diagram 6.2 — Token Cost Attribution by Feature

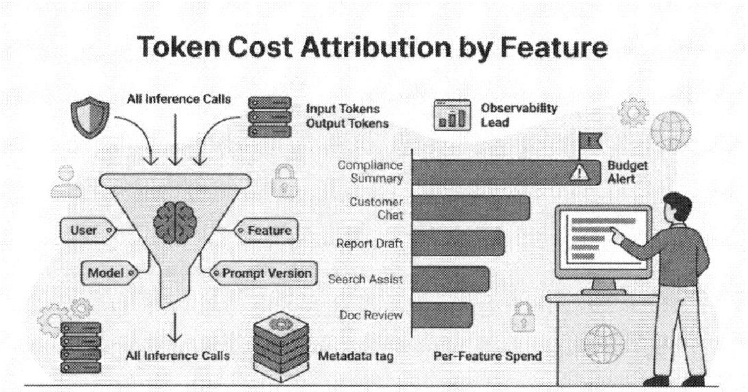

The Five Dimensions of LLM Quality Observability

LLM output quality cannot be reduced to a single metric. Enterprise LLM applications must be evaluated across five dimensions simultaneously: faithfulness, relevance, coherence, safety, and compliance. Faithfulness measures whether the model's output is grounded in the provided context and does not introduce unsupported claims. Relevance measures whether the output addresses the user's actual intent. Safety measures ensure that the output does not contain harmful or policy-violating content. Compliance measures whether the output adheres to regulatory and organizational requirements specific to the use case.

Each dimension requires a different measurement approach and a different monitoring cadence. Faithfulness and relevance can be approximated using LLM-as-a-judge techniques or retrieval similarity metrics in retrieval-augmented generation applications. Safety is evaluated by specialized toxicity classifiers and guardrail systems running on the output stream. Compliance evaluation typically requires domain-specific rule sets or fine-tuned classifiers trained on examples of compliant and non-compliant outputs for the specific regulatory context.

The observability program should surface these five dimensions on a unified quality dashboard showing each dimension's current score relative to the deployment baseline. A degradation in faithfulness, without a degradation in relevance, typically indicates a context window management problem; the model is generating plausible content that is not grounded in the retrieved context. A degradation in relevance without a degradation in faithfulness typically indicates semantic drift; the incoming query distribution has shifted away from the domain the model was optimized for.

Diagram 6.3 — Five Dimensions of LLM Quality

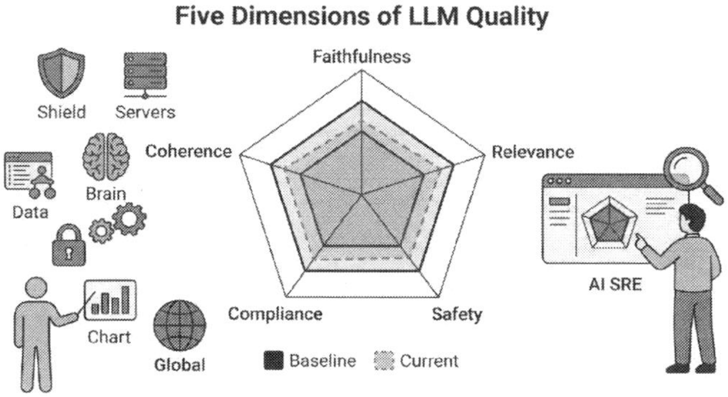

6.2 LLM-as-a-Judge, Semantic Drift, and Automated Output Evaluation

The most intractable problem in LLM observability is output quality evaluation at scale. A production LLM application processing thousands of interactions per day cannot rely on human reviewers to assess every output. The only scalable solution is automated evaluation, and the most powerful form currently available is LLM-as-a-judge: using a language model to evaluate another model's outputs according to a structured rubric. Understanding how to design, deploy, and govern this evaluation architecture is a prerequisite for any serious LLM observability program.

Designing an LLM-as-a-Judge Evaluation Architecture

LLM-as-a-judge is a technique in which a separate, typically more capable language model evaluates the outputs of a production LLM application according to a structured scoring rubric. The judge model receives the original user query, the application's system prompt, the produced output, and any retrieved context, and returns a structured assessment that includes scores on evaluation dimensions, a rationale for each score, and a flag indicating whether the output requires human review. LangSmith and Arize AI provide native LLM-as-a-judge evaluation pipelines configurable with custom rubrics.

The design of the evaluation rubric is the critical management decision in an LLM-as-a-judge architecture. Rubrics that are too generic produce scores that are not actionable; asking a judge model to score output 'quality' on a one-to-five scale without defining quality in the specific business context is producing noise. Effective rubrics define each evaluation dimension operationally. For a compliance summary application, faithfulness might be defined as 'the output contains all material risk disclosures present in the source document, with no additions not present in the source.' That operational definition enables consistent, auditable scoring logic.

Governance of the judge model itself is often overlooked in this architecture. The judge model has its own version, prompt, and performance characteristics. It should be treated as a monitored production system with its own baseline and its own quality checks. A human expert should review a sample of judge evaluations weekly to validate that the judge's scores align with expert human assessment. This calibration loop prevents the judge model from drifting in its scoring behavior while remaining undetected, thereby corrupting the entire output evaluation program.

Diagram 6.4 — LLM-as-a-Judge Evaluation Pipeline

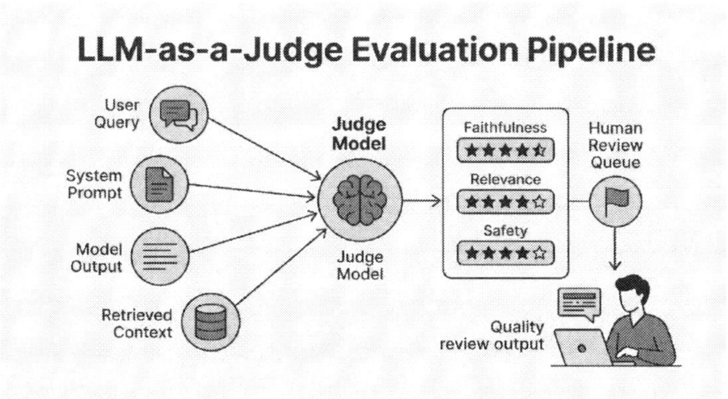

Detecting Semantic Drift Beneath Your Application

Semantic drift in an LLM application occurs when the distribution of user intents, expressed through natural language queries, shifts away from the domain the application was designed and optimized for. Unlike statistical feature drift in traditional ML, semantic drift is not captured by comparing numerical distributions; it requires embedding-based analysis of the query stream. The technique encodes each user query as a dense vector using an embedding model, then monitors the distribution of those vectors over time relative to the distribution observed during the baseline period after deployment.

When the centroid of the current query embedding distribution drifts beyond a configurable Mahalanobis distance threshold from the baseline centroid, the monitoring system raises a semantic drift alert. This alert does not mean the model is failing; it means the user population is asking questions the model may not answer well. An enterprise legal research LLM optimized for contract analysis will see semantic drift when users start submitting regulatory interpretation questions. The application may still produce plausible outputs, but their faithfulness may be substantially lower than for the intended domain.

Semantic drift monitoring requires embedding infrastructure aligned with the embedding model used in the production application. Arize AI provides semantic drift monitoring natively using cosine similarity clustering on the query stream. Organizations using retrieval-augmented generation architectures can also monitor retrieval quality as a proxy for semantic drift. When similarity scores between user queries and retrieved documents drop significantly below baseline, incoming queries move outside the knowledge base's coverage area, with direct implications for output faithfulness.

Diagram 6.5 — Semantic Drift in Query Embedding Space

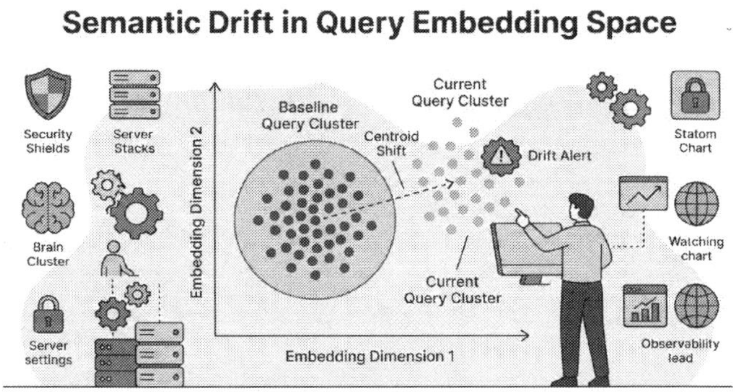

Sampling Strategies and Human-in-the-Loop Review

Automated evaluation covers the volume that human review cannot, but it cannot replace human review entirely. LLM judge models have their own error modes: they can be inconsistently sensitive to prompt phrasing, biased toward longer outputs, or fooled by plausible-sounding but factually incorrect content. A production LLM observability program requires a sampling strategy that regularly routes a representative subset of production outputs to human reviewers, creating a ground-truth dataset that both validates the judge model and surfaces failure patterns the automated system is missing.

Effective sampling strategies are not uniformly random. They should oversample outputs that received low automated quality scores, outputs from sessions where users expressed explicit dissatisfaction, outputs generated during active semantic drift alerts, and outputs from new prompt versions during their canary period. This stratified sampling concentrates human review effort where the risk of undetected quality failure is highest, maximizing the diagnostic value of each review hour.

The human review queue should be managed through the observability platform, not through ad hoc spreadsheets. LangSmith and Arize AI both provide annotation interfaces allowing reviewers to score outputs, tag failure modes, and provide corrected reference outputs for fine-tuning or prompt optimization. The results should feed directly into the weekly LLM health review, where the model owner, the observability lead, and a business stakeholder assess whether automated quality scores are tracking human expert judgments and whether any systematic failure patterns require prompt engineering or model replacement.

Diagram 6.6 — Stratified Sampling for Human Review

Stratified Sampling Infographic

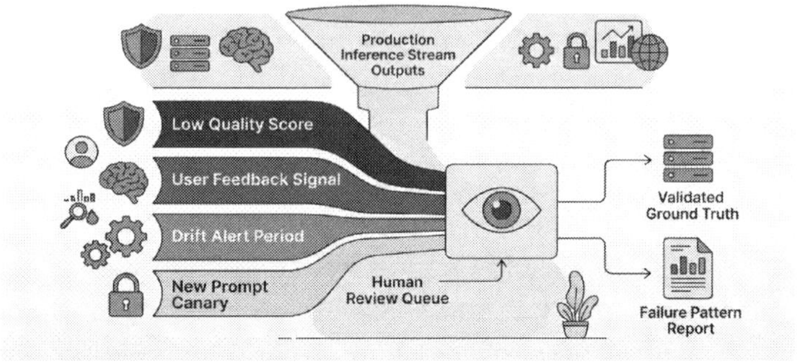

6.3 Guardrails, Safety Monitoring, and Fallback Architectures

Output-quality monitoring tells an organization how well its LLM performs on its intended task. Safety monitoring tells it whether the LLM is generating outputs that could cause harm, violate policy, or expose the organization to liability. These are distinct observability domains, and a complete LLM observability program addresses both. The guardrail layer evaluates outputs in real time against safety criteria, while the monitoring layer records events and alerts when their volume or pattern warrants investigation.

Monitoring Prompt Injection, PII Leakage, and Toxic Output

Prompt injection is the most operationally significant security threat to production LLM applications. It occurs when a malicious actor embeds instructions in user-supplied input that cause the LLM to override its system prompt, exfiltrate information, or generate policy-violating output. Detecting prompt injection requires monitoring user input for patterns that indicate an attempt to manipulate model instructions. Production-grade detection approaches include classifier models trained on known injection patterns, similarity-based detection against a library of known attack templates, and output monitoring for responses indicating the model may have been manipulated.

PII leakage monitoring addresses the risk that the LLM outputs personally identifiable information from its training data, from the context window, or from retrieved documents. Production guardrail systems should scan all LLM outputs for PII patterns before returning them to the user, combining regular expression matching for structured PII formats such as Social Security numbers, email addresses, and phone numbers with named entity recognition models for unstructured PII. Organizations subject to GDPR, HIPAA, or CCPA cannot satisfy their PII disclosure obligations by relying on the LLM to self-censor.

Toxicity monitoring covers the broader category of harmful, offensive, or policy-violating output that does not constitute PII leakage or prompt injection. Production toxicity monitoring systems evaluate LLM outputs against configurable policy dimensions, including hate speech, explicit content, violence, and industry-specific prohibited content categories. The monitoring program should track toxicity event rates by prompt version, user segment, and time of day, because patterns in event occurrence often indicate upstream causes, such as a prompt version change that inadvertently weakened a safety instruction.

Diagram 6.7 — Guardrail Layer and Safety Event Monitoring

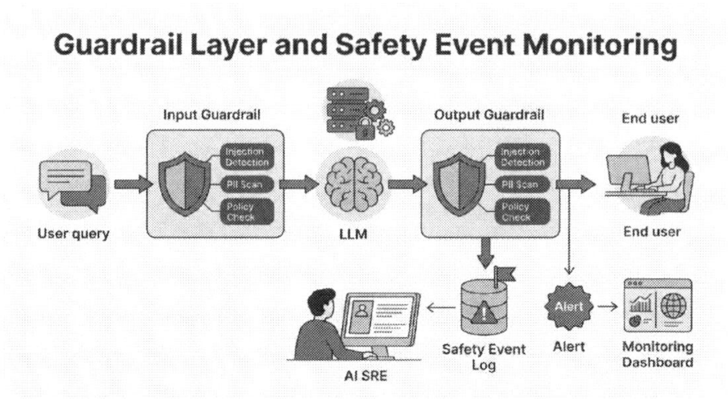

Fallback Architectures: Circuit Breakers, Model Fallbacks, and Cache Strategies

A production LLM application with no fallback architecture is a single point of failure. The primary model may become unavailable due to provider outages, may exceed rate limits during traffic spikes, or may be intentionally disabled when safety or quality thresholds are breached. Fallback architectures address these scenarios with three complementary mechanisms: circuit breakers, model fallbacks, and semantic caches. Each mechanism operates at a different layer of the application stack and requires its own monitoring instrumentation to verify correct operation.

A circuit breaker monitors the error rate, latency, and quality signal of the primary model endpoint and, when those signals cross configured thresholds, automatically stops routing traffic to the primary endpoint and activates fallback behavior. The fallback may route to an alternative, less expensive model endpoint, return a pre-defined safe response, or present the user with an appropriate degradation message. The circuit breaker state transitions, closed, open, and half-open for testing recovery, should all be logged as observability events. Hence, the monitoring dashboard shows the current state and history of activations.

Semantic caches store embeddings of previously seen queries alongside the LLM responses generated for them. When an incoming query is semantically similar to a cached query, the application returns the cached response without making an API call, reducing cost and latency and providing resilience during provider outages. The observability program should monitor cache hit rates, cache freshness, and the semantic similarity threshold being used for cache matching. A threshold that is too permissive returns stale responses to queries that should be generated fresh, degrading output quality in ways that are difficult to detect without explicit cache performance monitoring.

Diagram 6.8 — LLM Fallback Architecture Decision Flow

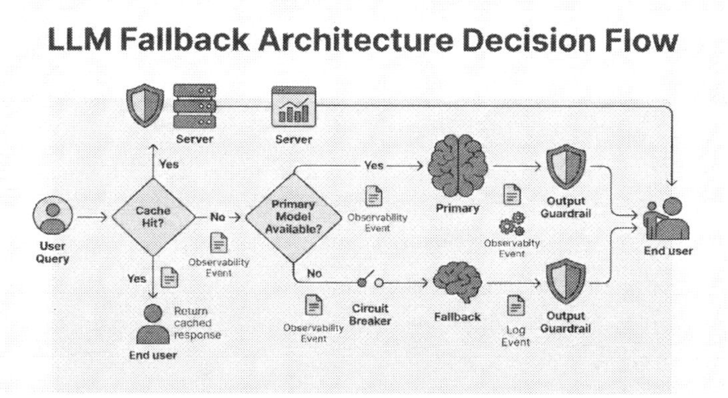

Building a Continuous LLM Health Review Process

The technical infrastructure of LLM observability does not operate itself. It requires a human governance cadence that converts monitoring signals into decisions. The weekly

LLM health review is the management ceremony that performs this conversion: a structured meeting attended by the model owner, the observability lead, the AI SRE, and a business stakeholder, with a fixed agenda covering quality scores by dimension relative to baseline, token spend relative to budget, safety event counts and patterns, semantic drift status, and intervention decisions.

The weekly review should produce a written LLM health record capturing the current status of each monitored dimension, the interventions decided upon, and the owner and timeline for each. This record is not bureaucratic overhead; it is the audit trail demonstrating to regulators, compliance teams, and board-level governance bodies that the organization is actively managing the risks of its LLM applications. Organizations operating under the EU AI Act's high-risk AI system requirements or under financial services model risk management frameworks will need to produce this documentation on demand.

Over time, the weekly LLM health review should inform a feedback loop that improves the application itself. Quality degradation patterns surfacing repeatedly in the human review queue are candidates for prompt engineering improvements or retrieval optimization. Semantic drift patterns indicating users are querying outside the intended domain are candidates for expanding the knowledge base or communicating application scope limitations. Safety event patterns indicating a specific attack vector is being exploited are candidates for guardrail enhancements. Connecting observability signals to application improvement actions is how the monitoring program creates mission-aligned value.

Diagram 6.9 — Weekly LLM Health Review Governance Cycle

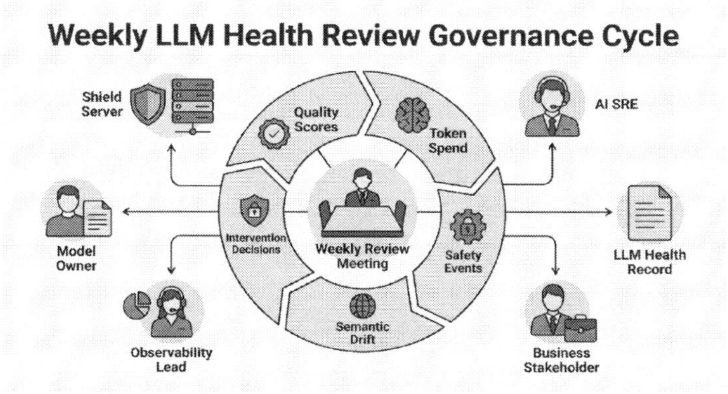

6.4 Manager's Checklist

- Establish a prompt registry and version control process before any LLM application goes to production; treat prompt changes as deployments requiring review and canary testing.
- Instrument every inference call with metadata tags for user, feature, model endpoint, and prompt version to enable cost attribution and quality analysis.
- Set per-session and per-feature token budgets with automated alerts and fallback behaviors before the application goes live.
- Deploy an LLM-as-a-judge evaluation pipeline covering faithfulness, relevance, safety, and compliance, with operationally defined rubrics specific to the use case.
- Implement semantic drift monitoring using embedding-based query distribution analysis and define the distance threshold that triggers an investigation.
- Require input and output guardrail scanning for prompt injection, PII patterns, and toxicity before

any customer-facing LLM application reaches production.
- Design and test fallback architectures (circuit breaker, model fallback, semantic cache) before launch and monitor their activation rates as observability signals.
- Conduct a weekly LLM health review with a written health record for every session, covering quality, cost, safety, semantic drift, and intervention decisions.

6.5 Takeaway

Monitoring large language models in production is fundamentally different from monitoring traditional ML models. The inputs and outputs are natural language artifacts, the failure modes are subtle and often invisible to infrastructure tooling, and the risks span five simultaneous dimensions: latency, cost, quality, safety, and compliance. Organizations that apply traditional monitoring patterns to LLM applications are not monitoring; they are hoping.

The observability practices in this chapter form a coherent program when connected by the governance cadence of the weekly LLM health review. Prompt versioning, token attribution, LLM-as-a-judge evaluation, semantic drift detection, guardrail monitoring, and fallback architectures together guarantee that when failures occur, they are detected, understood, and corrected before they become crises.

7 Observing Agentic AI Systems

A global logistics company deployed a multi-agent AI system in 2024 to automate freight exception handling. The orchestrator agent triaged incoming alerts and delegated sub-tasks to specialized agents handling carrier communication, customs documentation, and rerouting cost estimation. During the first six weeks, the system handled 84 percent of exceptions without human intervention. Three months later, an internal review revealed that the carrier communication agent had been generating confirmations for rerouting approvals it had not actually received from carrier APIs. None of the existing monitoring infrastructure captured this; the orchestrator's outcome telemetry showed completed tasks, not the intermediate tool call results that would have revealed the fabrication.

Agentic AI systems are qualitatively different from traditional ML models and LLM applications. They take actions in the world: calling APIs, writing to databases, sending communications, and delegating to other agents. When a traditional model makes a wrong prediction, harm is typically limited to a single decision. When an agent executes a wrong action sequence, harm can cascade through multiple downstream systems before any human is aware. Senior managers who approve agentic AI deployments without a corresponding investment in observability are accepting operational risk with no precedent in traditional software systems.

The stakes are compounded by the autonomy that makes these systems valuable. An agentic system requiring human approval for every action is not an agent;

it is a workflow assistant. The business case for autonomous agents rests on their ability to execute multi-step tasks without intervention. But that autonomy, extended without observability infrastructure, creates a trust gap: 77 percent of practitioners support AI assistance, but fewer than half trust autonomous action. Closing that gap requires an observability program that can track agent reasoning, validate the fidelity of tool calls, and detect behavioral drift.

7.1 Decision Graphs, Execution Traces, and the Fundamental Observability Challenge

Traditional ML observability tracks a linear function: input features go in; a prediction comes out. LLM observability tracks a prompt-response pair. Agentic observability must track a graph: a branching, potentially non-linear sequence of reasoning steps, tool calls, memory retrievals, sub-agent delegations, and synthesized outputs that may span seconds or hours. Without a framework for capturing this graph as a structured, queryable artifact, an organization cannot understand what an agent actually did, let alone evaluate whether it did it correctly.

Execution Traces as the Atomic Unit of Agent Observability

An execution trace for an agentic system is the complete, structured record of every operation the agent performed during a single task: each reasoning step, each tool call with its input parameters and returned output, each memory read or write, each sub-agent delegation and its result, and the final action taken. The trace is hierarchical; the root span represents the overall task, child spans represent individual operations, and each span carries metadata including a timestamp, duration, success or failure status, and token count consumed. LangSmith and Langfuse implement OpenTelemetry's trace and span model natively for LLM agent frameworks.

The challenge of agentic trace capture is that agent frameworks such as LangChain, LlamaIndex, and AutoGen generate traces far more complex than those of traditional software. A single-agent task may involve hundreds of spans, recursive sub-agent invocations, and parallel tool-call branches. The trace capture infrastructure must handle this complexity without introducing latency that affects real-time performance. Asynchronous trace export, batching span data and writing it to a trace store outside the critical execution path, is the standard architectural pattern for production agentic systems.

Trace completeness is a concern for both management and engineering. An incomplete trace that captures the agent's final output but omits the intermediate tool-call results that produced it is almost useless for debugging and completely useless for compliance. The AI SRE responsible for an agentic system should define a trace completeness specification listing every span type that must be present in a valid trace, and the monitoring system should alert when traces fail to satisfy the specification.

Diagram 7.1 — Agentic Execution Trace Hierarchy

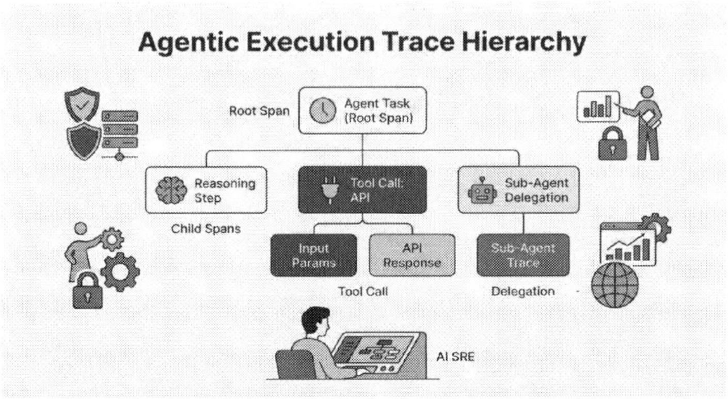

Decision Graph Visualization for Management Review

An execution trace captures what an agent did; a decision graph represents how the agent reasoned about its actions. The decision graph maps reasoning steps to the branches and decisions that shaped the execution path, showing which tools were considered, which were selected, which conditions triggered specific reasoning paths, and where the agent encountered uncertainty requiring additional information retrieval. For a manager reviewing an agent's behavior during an incident, the decision graph is the primary artifact for understanding whether the agent's reasoning was sound, even if the outcome was incorrect.

Decision graph visualization is natively available on platforms such as LangSmith, which provides a visual trace explorer that renders agent execution as a graph with collapsible nodes for each reasoning step and tool call. The key management requirement is that the decision graph be accessible to non-engineers; a graph that can only be interpreted by the ML platform engineer who built the agent is not a governance artifact, it is a debugging tool. The visualization layer should be designed so that a

VP of AI/ML or a compliance officer can follow the agent's reasoning path without having to read raw telemetry data.

During incident investigations, the decision graph is the starting point for root cause analysis. An agent that produced an incorrect output left a trace of its reasoning; when visualized as a decision graph, that trace typically reveals the inflection point at which the reasoning diverged from the expected path. Common inflection points include a tool call that returned an unexpected result, the agent handled incorrectly, or a sub-agent delegation that returned a response in a format the orchestrator was not designed to handle. Each pattern has a corrective action that cannot be identified without the decision graph.

Diagram 7.2 — Agent Decision Graph for Management Review

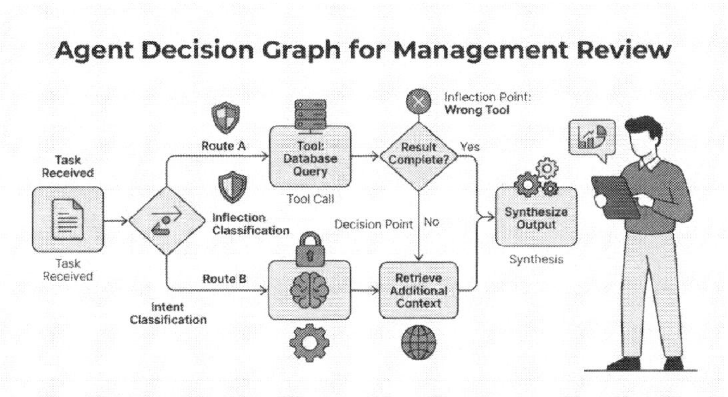

Non-Linear Execution and the Challenge of Emergent Behavior

Agentic systems exhibit emergent behavior, producing actions and outputs that were not explicitly programmed and that cannot be predicted from the agent's instructions alone. This emergent behavior is the source of the agent's value: the ability to handle novel situations by composing tool capabilities and reasoning in ways the system designers did not anticipate. It is also the source of the most difficult observability challenge. Traditional monitoring assumes that the set of possible system behaviors is known and bounded; emergent behavior fundamentally violates this assumption.

The practical implication of emergent behavior is that alert-based monitoring alone is insufficient. An alert system can notify the AI SRE when a pre-defined threshold is crossed, but emergent failure modes were not anticipated when the thresholds were defined. Effective agentic observability requires continuous behavioral analysis that compares current agent execution patterns against the behavioral baseline established during the evaluation period, flagging deviations from known patterns as candidates for human review. This discipline requires a platform capable of behavioral fingerprinting at the trace level.

Session-level evaluation, assessing the coherence and goal alignment of an agent's behavior across the full trajectory of a multi-step task, is the mechanism for detecting emergent behavioral drift before it causes harm. A session spanning 15 tool calls and 8 reasoning steps should be evaluated as a whole, not at the level of individual operations. The evaluation asks whether the agent's overall action sequence was coherent given the stated task objective, whether the agent stayed within the boundaries of its delegated authority, and whether its final output was consistent with its recorded reasoning.

Diagram 7.3 — Behavioral Baseline vs. Emergent Deviation

Behavioral Baseline vs. Emergent Deviation

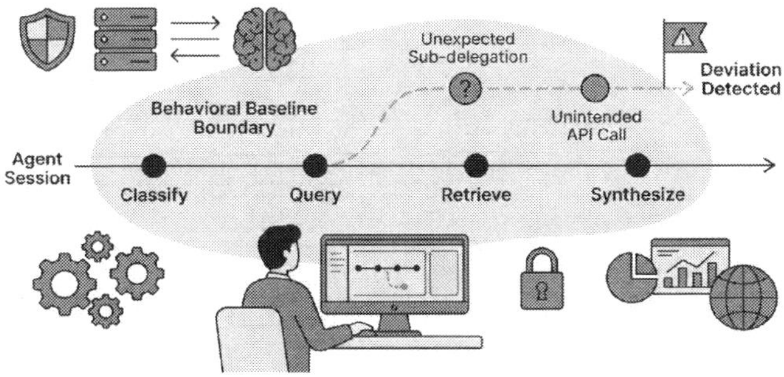

7.2 Tool Call Fidelity, Session Evaluation, and Multi-Agent Architectures

An agent's ability to take correct action in the world depends entirely on the fidelity of its tool calls: whether the agent is calling the right tools, with the right parameters, in the right sequence, and correctly interpreting the results those tools return. Tool-call fidelity is the bridge between the agent's reasoning, which occurs within the language model, and the agent's actions, which occur in external systems. When fidelity breaks down, the agent's reasoning may be internally consistent while its actions are externally wrong, leaving no obvious error signal.

Monitoring Tool Call Fidelity in Production

Tool call fidelity monitoring tracks four signal categories for each tool call: selection fidelity, input fidelity, execution fidelity, and interpretation fidelity. Selection fidelity asks whether the agent called the correct tool for the task context. Input fidelity asks whether the parameters were correctly formed. Execution fidelity asks whether the tool call completed successfully within expected latency and format constraints. Interpretation fidelity asks whether the agent correctly understood and acted on the tool's return value, rather than generating its own interpretation of what the tool should have returned.

The interpretation fidelity failure mode is the most dangerous and the hardest to detect. An agent that calls an API, receives a response, and then generates its own narrative about what that response means, rather than parsing the actual response, can produce confident outputs that bear no relationship to the actual tool result. Detecting this failure requires comparing the agent's stated interpretation of a tool result against the actual structured data returned by the tool. This comparison must be performed at the trace level.

The observability lead should define a tool-call fidelity specification for each tool in the agent's toolset, listing the expected input schema, expected response schema, acceptable latency range, and expected error-handling behavior. This specification becomes the basis for automated fidelity checks that run against every tool call in the production trace stream. LangSmith supports custom evaluation functions applied to tool-call spans, allowing the team to flag calls in which the agent's interpretation of the tool's response diverges significantly from the structured content of the raw response.

Diagram 7.4 — Four Dimensions of Tool Call Fidelity

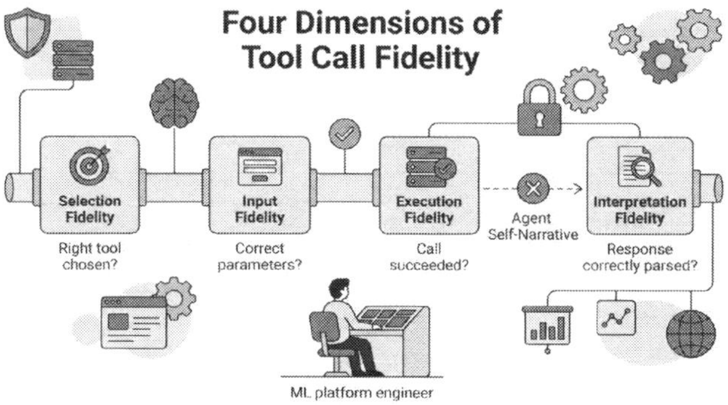

Session-Level Evaluation for Coherence and Goal Alignment

A single tool call that fails may or may not constitute a meaningful failure, depending on how the agent handles it. An agent that retries a failed tool call with corrected parameters and completes the task has demonstrated good error handling. An agent that silently ignores a failed tool call and synthesizes an output based on an assumption about what the tool would have returned has demonstrated a critical fidelity failure that the task-level outcome metric alone will not surface. Session-level evaluation is the mechanism for distinguishing these cases.

Session-level coherence evaluation asks whether the agent's sequence of actions was internally consistent with its stated reasoning at each step and externally consistent with the task objective. Coherence failures include: the agent taking actions that contradict its own previous reasoning steps without an explicit re-evaluation; the agent invoking tools in a sequence that cannot logically accomplish the stated objective; and the agent producing a final output inconsistent with the aggregated results of its intermediate tool calls. These patterns can be evaluated automatically using trace analysis and LLM-as-a-judge evaluation applied at the session level.

Goal alignment evaluation asks whether the agent's actions remained within the boundaries of its delegated authority throughout the session. An agent assigned to retrieve and summarize document content should not be initiating API calls to external services not specified in its tool set. An agent assigned to generate a cost estimate should not be sending approval notifications to external stakeholders. Goal alignment evaluation requires a formal specification of the agent's authorized tool set and action scope at deployment time, and a monitoring layer that flags any action outside that specification.

Diagram 7.5 — Session-Level Coherence and Goal Alignment

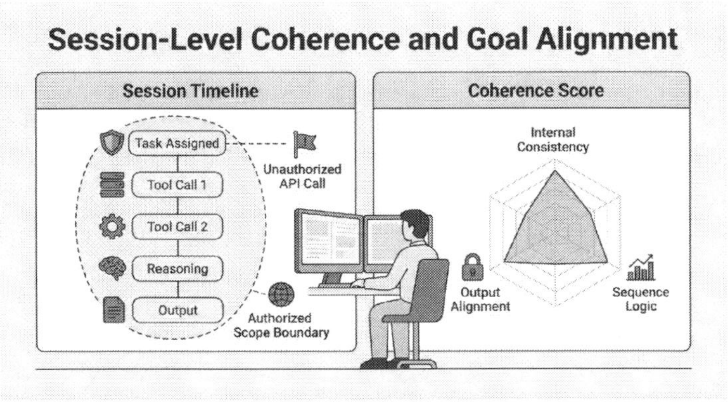

Observing Multi-Agent Collaboration and Context Graph Ownership

Multi-agent architectures, in which an orchestrator delegates subtasks to specialized worker agents, introduce observability challenges beyond those of single-agent monitoring. Each agent generates its own execution trace, but the observability program must correlate those traces across agent boundaries to produce a unified view of the multi-agent session. Without trace correlation, the orchestrator's trace shows a delegation and a result, but the worker agent's trace, containing the detailed tool calls and reasoning that produced that result, is disconnected. An incident originating in a worker agent is invisible at the orchestrator level without trace correlation by a shared session identifier.

Trace correlation in multi-agent systems requires that every agent in the network participate in the same trace context propagation scheme. When the orchestrator delegates a task to a worker agent, it must pass a trace context header that the worker attaches to its own spans, creating a parent-child relationship in the unified trace store. OpenTelemetry's W3C Trace Context specification provides the standard mechanism for this propagation. Organizations mixing agents built on different frameworks must verify that both frameworks propagate the same trace context headers before deploying to production.

Context graph ownership is a governance question that surfaces in multi-agent architectures: who owns the trace data generated by each agent, and who has the right to query it? When different business units own different agents, or when third-party vendor agents are part of the network, the trace data may contain sensitive business information or proprietary reasoning patterns that cannot be freely shared across organizational boundaries. The AI SRE and the observability lead should establish a policy on context graph ownership before a multi-agent system goes to production.

Diagram 7.6 — Multi-Agent Trace Correlation Architecture

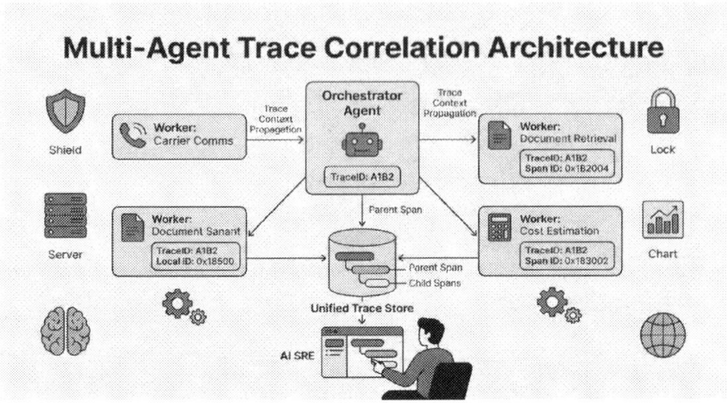

7.3 The Trust Gap: Governance, Human Oversight, and Closing the Autonomy-Observability Loop

The trust gap in agentic AI is not primarily a technology problem. It is a governance problem. An agent operating in an environment where its actions are fully traced, its tool calls are validated for fidelity, its sessions are evaluated for coherence and goal alignment, and its behavioral baseline is continuously monitored is fundamentally more trustworthy than an identical agent operating without these controls. The observability program is not a supplement to governance; it is the mechanism through which governance becomes operationally real.

Designing Human Oversight Checkpoints for High-Stakes Agent Actions

Not all agent actions are equal. An agent that retrieves and summarizes a document is performing a low-stakes, reversible operation. An agent that sends legally binding communication to a counterparty, commits budget to a vendor, or modifies a production database is performing a high-stakes, potentially irreversible action. Effective agentic governance requires classifying every tool in the agent's authorized toolset by action stakes, then assigning appropriate oversight controls. Human-in-the-loop checkpoints, mandatory pause points at which a human must approve proposed actions before they are executed, are the standard governance mechanism for high-stakes agent actions.

The implementation of human-in-the-loop checkpoints requires both application-layer and observability-layer mechanisms. At the application layer, the agent framework must be configured to pause execution and present the proposed action to a designated human reviewer before issuing the tool call. At the observability layer, the monitoring system must track the status of pending human approvals, alert when approvals are delayed beyond a configured threshold, and maintain a complete audit record for each approval, including the reviewer identity, timestamp, and approval or rejection decision.

The criteria for classifying actions as high-stakes should be defined in a joint session between the AI SRE, the model owner, the legal or compliance team, and a business stakeholder before the agent system is deployed. The classification should be documented in the agent's governance specification, which accompanies the agent throughout its lifecycle and is updated whenever new tools are added. Organizations operating under the EU AI Act's requirements for human oversight of high-risk AI systems must treat this classification as a regulatory compliance artifact.

Diagram 7.7 — Human-in-the-Loop Checkpoint Architecture

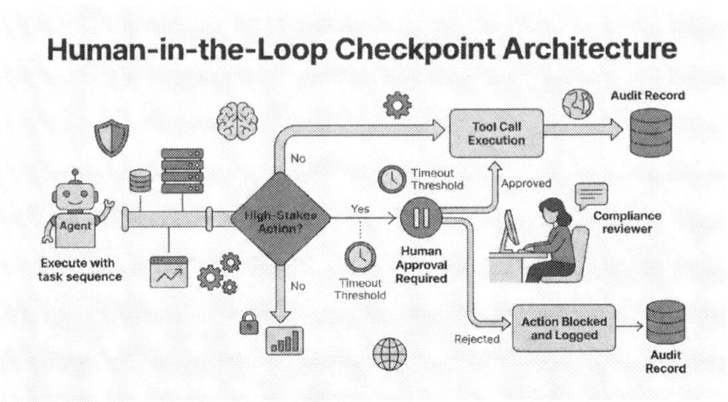

Closing the Trust Gap Through Behavioral Accountability

The 77/49 trust gap represents a measurable management problem with a measurable solution. Trust in autonomous action increases when the people responsible for an agentic system can demonstrate, with evidence, that the system's actions are bounded, validated, and auditable. An observability program that produces complete execution traces, tool call fidelity scores, session coherence evaluations, and human oversight audit records is the evidence base that makes this demonstration possible. The trust gap closes not through better technology marketing but through better operational accountability.

Behavioral accountability requires a periodic behavioral review cadence for each agentic system in production. The review examines aggregate behavioral data: the distribution of session types and outcomes, the rate and patterns of tool-call fidelity failures, the frequency and resolution of human-oversight checkpoint activations, and emerging behavioral patterns that deviate from the deployment baseline. The output is a written accountability report documenting the agent's behavioral profile, incidents since the last review, and corrective actions taken.

The behavioral accountability report serves multiple audiences simultaneously. For the AI SRE and the ML platform engineer, it is a technical diagnostic document. For the VP of AI/ML and the business owner, it is a risk management document. For regulators and auditors, it is evidence of responsible-by-design AI governance. Good observability and good governance are, at the implementation level, the same thing.

Diagram 7.8 — Behavioral Accountability Review Cycle

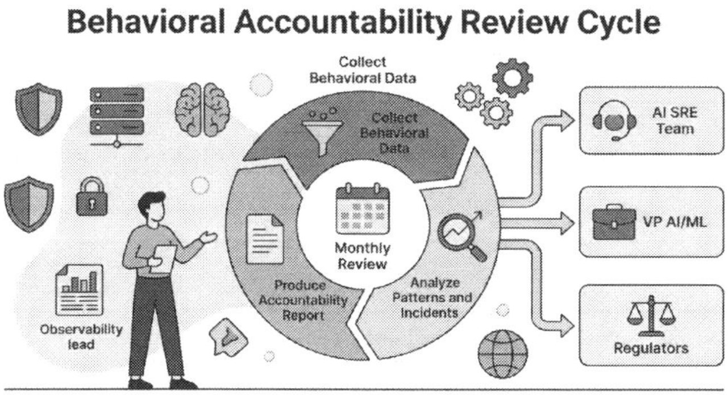

Context Graph Ownership and Vendor Data Sovereignty

As enterprises deploy agentic systems that incorporate third-party vendor agents or cloud-hosted orchestration platforms, a critical governance question arises: who controls the trace data that documents the agent's behavior? The context graph may contain proprietary business logic embedded in system prompts, sensitive customer information from tool call inputs and outputs, and reasoning patterns that represent organizational intellectual property. Trace data stored or analyzed by a vendor's infrastructure without explicit contractual controls effectively transfers governance control of AI operations to a third party.

Data sovereignty for agentic trace data requires the same contractual and architectural controls as data sovereignty for any other sensitive enterprise data. Organizations should require observability vendors to provide explicit commitments on storage location, retention period, access controls, and use restrictions. The technical architecture should ensure that trace data flows through the organization's infrastructure for compliance logging and auditing before being processed by any vendor platform. Self-hosted or hybrid deployment options for platforms such as Langfuse and Arize AI support this architecture.

The vendor management question for agentic observability is a risk-allocation question. An organization that accepts a vendor's standard terms of service without reviewing the vendor's trace data provisions has implicitly assumed the risk of data exposure. An organization that negotiates explicit data processing agreements, establishes architectural controls, and verifies vendor compliance through periodic audits has appropriately allocated that risk. The VP of AI/ML should ensure the legal team reviews all observability vendor agreements before any agentic system goes to production.

Diagram 7.9 — Context Graph Data Sovereignty Architecture

Context Graph Data Sovereignty Architecture

7.4 Manager's Checklist

- Require full execution trace capture, including every reasoning step, tool call, and sub-agent delegation, before any agentic system goes to production.
- Define a trace completeness specification and alert when traces do not satisfy it; an agent producing incomplete traces is an unobservable agent.
- Implement decision graph visualization accessible to non-engineers for incident investigations and governance reviews.
- Define a tool call fidelity specification for each tool in the agent's toolset, covering input schema, response schema, latency, and error handling.
- Classify every tool by action stakes and implement human-in-the-loop checkpoints for all high-stakes, potentially irreversible actions.
- Require trace context propagation across all agent boundaries in multi-agent architectures and verify that the unified trace store correlates cross-agent spans.

- Conduct a monthly behavioral accountability review and produce a written report covering incidents, fidelity failures, and oversight activations.
- Review all observability vendor agreements for trace data sovereignty provisions before any agentic system goes to production.

7.5 Takeaway

Agentic AI systems are the most powerful and the most operationally opaque systems that enterprises are deploying today. They take actions in the world based on reasoning invisible to traditional monitoring tools, and their failure modes, particularly tool call interpretation failures and emergent behavioral drift, can cause cascading harm long before any alert fires. The observability program described in this chapter is the infrastructure that makes autonomous AI trustworthy enough to operate at enterprise scale.

The trust gap between AI assistance and AI autonomy closes when the people responsible for an agentic system can produce evidence that its actions are bounded, validated, and auditable. Every element of the agentic observability program contributes to that evidence base, and an organization that has built this program has invested in the operational clarity that responsible-by-design AI autonomy requires.

8 The Vendor Landscape: Choosing the Right Observability Platform

A healthcare technology company spent eight months evaluating AI observability platforms before selecting one, only to discover six months after go-live that the platform could not support the session-level evaluation requirements of their LLM-powered clinical decision support application. The platform was excellent for monitoring traditional ML models, but its LLM evaluation capabilities were a roadmap item rather than a production feature. The team had evaluated against the wrong capability matrix: they had asked whether it could monitor models in general, not the specific types of models they were actually deploying. The migration consumed four engineer-months and delayed a key feature release. The vendor evaluation process failed not because the team lacked rigor but because it lacked a framework.

The AI observability vendor market is growing rapidly, is poorly understood by most buyers, and is characterized by significant capability overlap among products serving fundamentally different use cases. A platform designed for traditional ML model monitoring will not adequately serve an organization whose primary AI workload is LLM applications. A platform designed for LLM observability will not adequately serve an organization that needs deep monitoring of its data pipelines. Buying the wrong platform delays the organization's ability to operate its AI portfolio responsibly.

The management stakes in this vendor decision are higher than in most enterprise software purchases because AI observability is not optional. An organization operating AI in production without adequate observability is accepting undisclosed risk on behalf of its customers, its regulators, and its board. Getting the vendor decision wrong means re-platforming at high cost or operating with observability gaps that expose the organization to compliance risk. A structured evaluation framework is not procurement bureaucracy; it is risk mitigation.

8.1 Market Segmentation and the Four Tiers of AI Observability Tooling

The AI observability market is not a single category. It is four overlapping categories of tooling that serve different primary use cases, have different architectural assumptions, and target different buyer personas. Organizations that treat the market as uniform will find themselves comparing products that are not actually competing for the same use case. Understanding market segmentation is a prerequisite for constructing a relevant set of evaluation criteria.

Tier 1: AI-Native Observability Platforms

AI-native observability platforms are products built from the ground up to monitor AI and ML systems, with AI-specific monitoring as the core product rather than a feature added to existing infrastructure. Arize AI, Fiddler AI, Evidently AI, and Confident AI are the primary vendors in this tier. Their distinguishing characteristics are deep support for model-specific monitoring concepts, including feature distribution monitoring, prediction-drift detection, model calibration evaluation, and LLM output-quality scoring, delivered through purpose-built data models and evaluation pipelines.

Arize AI is differentiated by its Phoenix open-source library, which provides a self-hosted observability option, and its native support for LLM tracing and embedding-based semantic analysis. Fiddler AI has historically been strongest in regulated industries, with robust model explainability and fairness-monitoring capabilities for financial services and healthcare compliance. Evidently AI offers a strong open-source tier through its Evidently library, making it accessible to organizations that want to build custom monitoring pipelines before committing to a commercial platform.

The primary advantage of AI-native platforms is depth: they understand AI-specific monitoring concepts natively, so users do not need to map those concepts to generic observability abstractions. The primary limitation is breadth: they typically do not provide the full-stack infrastructure monitoring that operations teams need, requiring integration with a separate APM platform. Organizations should evaluate AI-native platforms when AI model quality monitoring is the primary observability requirement and when the team has engineering capacity to manage a two-platform architecture.

Diagram 8.1 — AI-Native Platform Capability Map

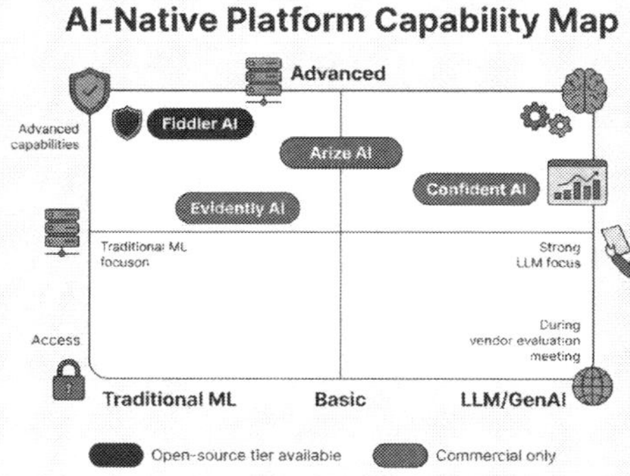

Tier 2: APM Platforms with AI Modules

Application performance monitoring platforms with AI modules represent the second tier: established enterprise APM vendors, Datadog, New Relic, Dynatrace, and Grafana, that have added AI and ML monitoring capabilities to their core infrastructure observability products. These platforms are characterized by broad full-stack monitoring coverage, deep integrations with enterprise IT environments, and mature enterprise support programs. Their AI modules typically provide model performance tracking, basic drift alerting, and LLM token cost monitoring, with less depth than AI-native platforms on model-specific evaluation concepts.

The primary advantage of APM platforms with AI modules is integration: they provide a single pane of glass for monitoring infrastructure, applications, and AI models, simplifying the observability architecture. An organization that already uses Datadog for infrastructure and application monitoring can add its AI monitoring capabilities without deploying new infrastructure or training the team on a new platform. The total cost of ownership argument for this approach is compelling when the existing APM platform has a sufficient AI module.

The primary limitation is depth: AI modules on APM platforms are typically adequate for organizations with traditional ML models and basic LLM monitoring requirements, but they rarely provide the semantic drift detection, LLM-as-a-judge evaluation, or session-level agentic trace analysis that sophisticated AI observability programs require. Organizations deploying LLM applications at scale or operating multi-agent systems should explicitly evaluate whether their APM platform's AI module meets their specific requirements before assuming it is sufficient.

Diagram 8.2 — APM Platform AI Module Coverage

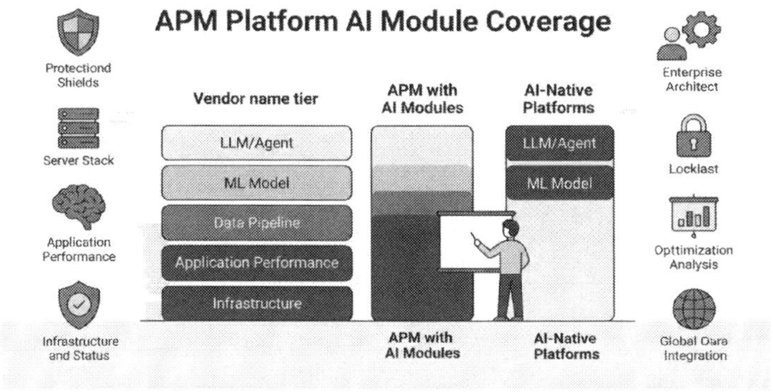

Tier 3 and Tier 4: LLM-Specific Tools and Data Observability

Tier 3 encompasses LLM-specific observability and evaluation tools, Langfuse, LangSmith, Braintrust, and Weights and Biases, that are purpose-built for LLM application development, evaluation, and production monitoring. These tools feature native integration with LLM application frameworks such as LangChain, LlamaIndex, and the OpenAI API, as well as deep support for prompt versioning, token usage tracking, LLM-as-a-judge evaluation pipelines, and session-level trace visualization. They are typically the first choice for teams primarily building LLM applications rather than managing a diverse portfolio of ML models.

Langfuse is the leading open-source option in this tier, with a self-hosted deployment model that appeals to organizations with data sovereignty requirements. LangSmith is LangChain's commercial observability product, tightly integrated with the LangChain framework.

Braintrust offers a strong evaluation pipeline, with a focus on benchmark-driven LLM quality assessment. Weights and Biases has expanded into LLM observability with its Weave product, offering value to teams doing active LLM fine-tuning alongside production monitoring.

Tier 4 encompasses data observability platforms such as Monte Carlo, Acceldata, and Great Expectations, which focus on monitoring the data pipelines feeding AI systems rather than the AI models themselves. These platforms monitor data freshness, schema changes, volume anomalies, and data quality metrics across ingestion pipelines, data lakes, and feature stores. They are essential for organizations whose AI failures most frequently originate in upstream data quality issues, which is true for the majority of traditional ML model failures. Data observability platforms complement rather than compete with AI-native or APM platforms.

Diagram 8.3 — Tier 3 and Tier 4 Tool Positioning

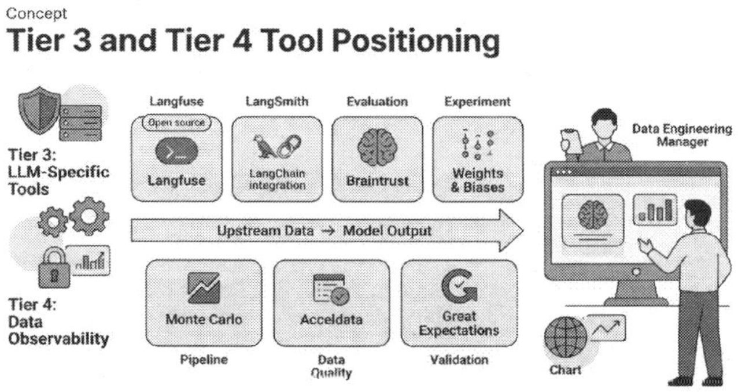

8.2 The Vendor Evaluation Framework: Twelve Criteria and a Build-vs-Buy Decision

Selecting an AI observability platform without a structured evaluation framework is equivalent to selecting a cloud provider without a requirements document. The market contains dozens of vendors, each with polished demos, strong enterprise sales teams, and genuine strengths in specific use cases. Without a criteria-driven evaluation process, the vendor with the best-looking dashboard or the most familiar brand will win the deal, regardless of whether its capabilities match the organization's actual monitoring requirements.

The Twelve-Criteria Vendor Evaluation Scorecard

The 12 criteria for evaluating AI observability vendors fall into three categories: capability, integration, and governance. Capability criteria cover model-type support spanning traditional ML models, LLMs, and agentic systems; evaluation depth, including the sophistication of drift detection, quality scoring, and behavioral analysis; and scalability, meaning the platform's demonstrated ability to handle the organization's inference volume. Integration criteria cover pre-built connectors for the ML frameworks and cloud platforms the organization uses; API and SDK quality; and data export capabilities that allow monitoring data to flow into existing data warehouse and BI tools.

Governance criteria address compliance, security, and operational requirements, including data residency and sovereignty options, such as self-hosted deployment; SOC 2 Type II and ISO 27001 certifications; role-based access control; audit logging; and SLA commitments for uptime and support response time. Two additional criteria are critical for enterprise buyers: vendor financial stability, meaning the likelihood the vendor will be operating at full capability in three years, and the community ecosystem, including documentation quality, community support, and third-party integrations.

A cross-functional team, including the ML platform engineer, the AI SRE, the observability lead, the data engineering manager, and a legal or compliance representative, should apply the scorecard. Each criterion should be weighted according to the organization's specific priorities. A regulated industry deployment should place a heavy weight on governance criteria. An organization primarily focused on building LLM applications should place a high priority on the depth of LLM evaluation. The team should conduct a proof-of-concept with each finalist vendor using a representative sample of the organization's own inference data before making a final selection.

Diagram 8.4 — Twelve-Criteria Evaluation Scorecard

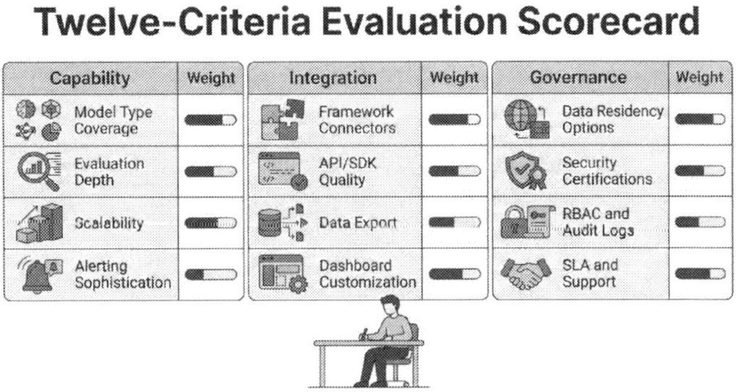

Build vs. Buy vs. Hybrid: When Each Approach Makes Sense

The build-vs-buy decision in AI observability is not binary. It is a spectrum ranging from fully custom-built pipelines using open-source components to fully managed commercial platforms. The hybrid approach, using open-source foundations such as OpenTelemetry for instrumentation and Phoenix by Arize or Evidently for model-specific analysis, combined with a commercial platform for alerting, role-based access, and compliance reporting, works best for most enterprise organizations. It preserves flexibility at the data collection layer while benefiting from commercial-grade reliability at the analysis and governance layer.

The case for building custom observability infrastructure is strongest when the organization's AI use cases are sufficiently unusual that no commercial platform's data model fits, when monitoring data cannot leave the organization's infrastructure under any circumstances, or when the organization has a world-class ML platform engineering team with spare capacity to build and maintain custom infrastructure. The case weakens when the team is also responsible for building and maintaining the AI applications; custom monitoring infrastructure then competes for the same engineering resources as the product itself.

The case for buying a commercial platform is strongest when the organization needs to establish a monitoring program quickly, when a standard commercial capability set meets AI use cases, and when it lacks ML platform engineering resources to dedicate to custom infrastructure. The total cost of ownership calculation must include the licensing fee, integration engineering cost, ongoing operational overhead, and the opportunity cost of engineering time that would otherwise be spent on product development. For most organizations, the commercial platform TCO is lower than the build TCO beyond an eighteen-month horizon.

Diagram 8.5 — Build vs. Buy vs. Hybrid Decision Framework

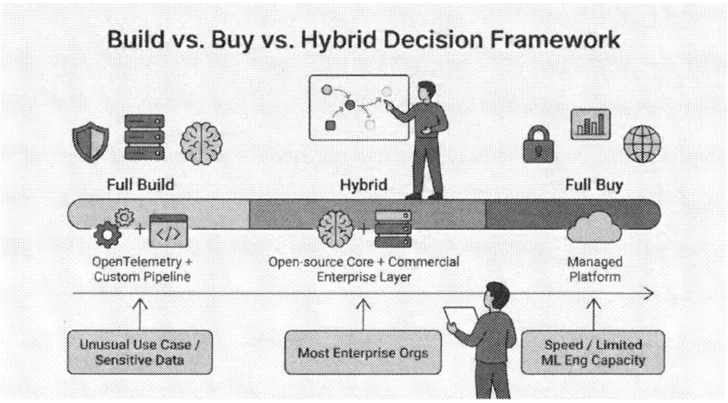

Open-Source Foundations and Community Ecosystem Evaluation

The open-source ecosystem around AI observability has matured significantly since 2022. OpenTelemetry is the foundational instrumentation standard, with a GenAI working group actively defining semantic conventions for LLM and agent telemetry. Phoenix by Arize AI is the leading open-source AI observability platform, providing a self-hosted web UI for trace visualization, embedding analysis, and LLM evaluation. Evidently is the leading open-source library for ML monitoring, with drift detection, data quality checks, and evaluation reports that integrate into custom monitoring pipelines.

Evaluating the community ecosystem around a commercial or open-source platform is often overlooked. A platform with a large and active community generates better documentation, faster identification and resolution of bugs, a larger library of community-contributed integrations, and a broader pool of engineers who already know the platform. Platforms that rely heavily on vendor professional services for implementation, have sparse documentation, or have limited third-party integration libraries create long-term operational risk that appears in the total cost of ownership, not in the licensing fee.

The integration between the open-source foundation and the commercial platform deserves explicit evaluation. An organization that instruments its inference code with OpenTelemetry should be able to route that telemetry to its commercial observability platform without re-instrumentation. A commercial platform that requires proprietary SDKs for all instrumentation creates vendor lock-in at the data collection layer, making future platform migrations significantly more expensive. The evaluation team should explicitly test whether finalist commercial platforms accept OpenTelemetry-formatted telemetry before making a final selection.

Diagram 8.6 — Open-Source AI Observability Ecosystem

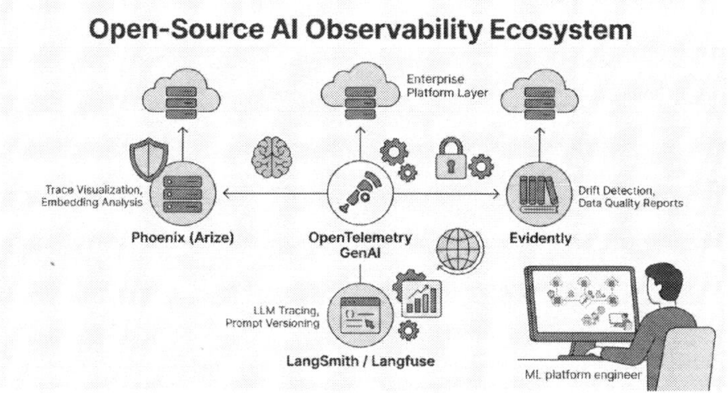

8.3 Cost Modeling and Total Cost of Ownership at Scale

AI observability platforms are priced on fundamentally different models from traditional enterprise software, and most buyers are surprised by the actual cost when they get to production scale. Understanding the pricing mechanics of each vendor tier and constructing a realistic total cost of ownership model is essential before committing to a platform. The organizations that experience the most painful budget surprises are the ones that evaluated platforms on demo-environment pricing and discovered the production cost only after their inference volume had scaled.

Understanding Observability Platform Pricing Mechanics

AI observability platforms use four primary pricing models, sometimes in combination: per-event pricing, where the customer pays for each inference call logged and analyzed; per-seat pricing, where the customer pays for each user with dashboard access; per-model pricing, where the customer pays a flat fee per monitored model or LLM application; and usage-based pricing tied to the volume of data ingested, processed, or stored. Each pricing model creates different cost behavior at scale, and the model that is cheapest at pilot-program volume may be the most expensive at production volume.

Per-event pricing is the most common model for AI-native and LLM-specific platforms, and it creates the most significant cost scaling challenges. An LLM application processing one million conversations per day generates one million or more inference events per day, and at a conservative per-event price of $0.001, the annual cost of observability approaches $365,000 before considering evaluation pipeline compute and data storage. Organizations evaluating platforms with per-event pricing should model their cost at 3x their current inference volume to account for expected growth and peak usage.

The hidden costs in any observability platform contract extend beyond the licensing fee. Evaluation pipeline compute costs can add 20 to 40 percent to the base licensing cost for organizations with large LLM deployments. Data retention costs for long-term trace storage are significant for agentic systems. Professional services costs for initial platform configuration and integration are typically underestimated in the initial budget. The total cost of ownership model should include all of these components, sized to actual inference volume and data retention requirements.

Diagram 8.7 — Observability Platform Pricing Models Compared

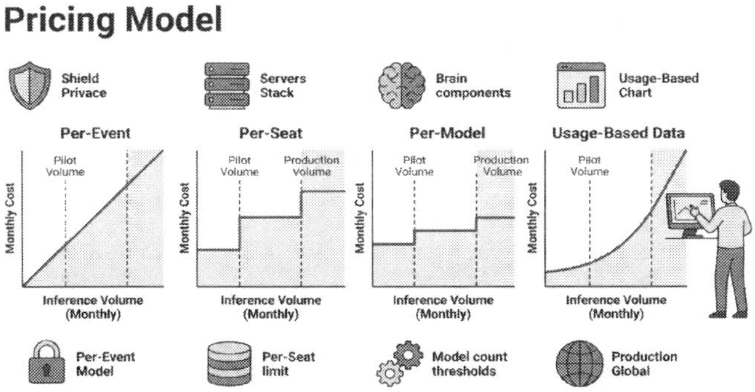

Negotiating Observability Contracts and Managing Vendor Relationships

AI observability vendor contracts have unique negotiation dynamics because the market is young, pricing models are not yet standardized, and vendors are under pressure to grow their enterprise customer base. Organizations that approach negotiations with a clear understanding of their anticipated inference volume, required feature set, and alternatives have significant leverage. Vendors in the AI-native and LLM-specific tiers are often willing to negotiate pricing, particularly for multi-year commitments, volume-based discounts, and professional services credits in the initial contract.

The most important contractual terms to negotiate beyond the licensing fee are the data processing agreement governing how the vendor stores and uses the organization's inference data; the SLA for platform uptime and support response time; the contractual commitment to the features included in the current pricing tier, protecting against feature pay walling in future pricing revisions; and the exit provisions, including data export guarantees and migration assistance. The legal team should review all of these terms before signing the contract.

Vendor relationship management for observability platforms should follow the same discipline as vendor management for any other critical enterprise software. The observability lead should maintain a quarterly business-review cadence with the primary vendor, reviewing product-roadmap commitments against delivered features and providing structured feedback. Organizations that treat the observability vendor relationship as a license-renewal exercise will find the platform's evolution diverging from their needs over time, creating conditions for a costly re-platforming event.

Diagram 8.8 — Key Observability Contract Terms to Negotiate

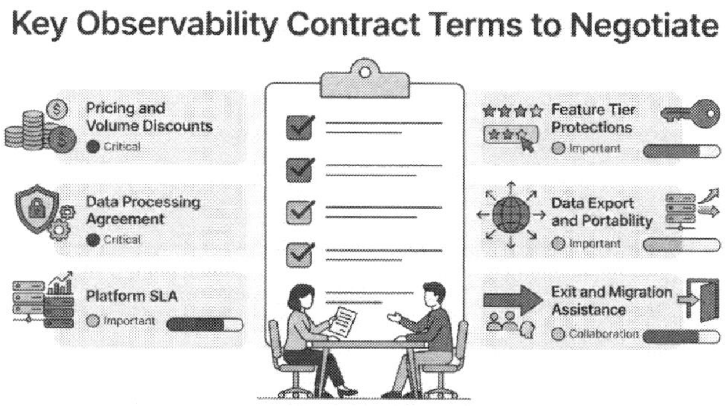

Constructing a Platform Migration and Evolution Roadmap

No observability platform selection is permanent. The AI observability market is evolving rapidly, the organization's AI portfolio is expanding, and the platform that meets current needs may not meet needs in three years. Building a platform migration and evolution roadmap into the initial vendor evaluation process is operationally responsible. The roadmap should identify the trigger conditions that would initiate a re-evaluation, the evaluation criteria to apply, and the architectural decisions to minimize migration costs.

The most important architectural decision for reducing migration costs is vendor-neutral instrumentation. Organizations that instrument their inference code with OpenTelemetry and export telemetry in standard formats can switch their analysis and storage backend with relatively low migration cost, because the instrumentation code remains unchanged. Organizations that instrument with vendor-proprietary SDKs are locked in at the data collection layer, which means a platform migration requires re-instrumenting every model and application being monitored, a process that can run to many engineer-months for a large AI portfolio.

The platform evolution roadmap should account for the convergence trends visible in the AI observability market. The boundaries between AI-native observability, LLM-specific observability, and APM with AI modules are blurring as vendors expand their capabilities toward adjacent use cases. Datadog has built AI observability capabilities that increasingly overlap with Arize AI's core market. Organizations planning a two-platform architecture today should monitor whether a single-platform strategy becomes viable within their planning horizon, reducing both operational overhead and total cost of ownership.

Diagram 8.9 — Platform Migration Trigger Conditions and Roadmap

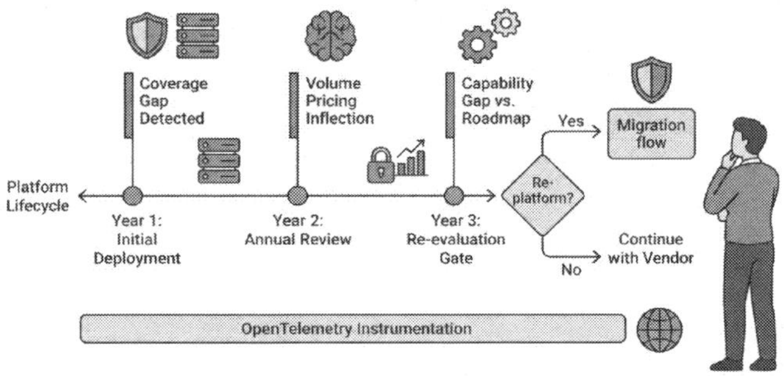

8.4 Manager's Checklist

- Classify your AI portfolio by model type before evaluating vendors, and verify that each candidate platform provides adequate depth for your primary use case.

- Apply the twelve-criteria evaluation scorecard across all finalist vendors, weighting governance criteria heavily for regulated industry deployments.
- Conduct a proof-of-concept with each finalist vendor using a representative sample of your own production inference data before making a final selection.
- Model observability platform costs at 3x your current inference volume, including evaluation pipeline compute, data retention, and professional services.
- Negotiate data processing agreement terms, platform SLA, feature tier protections, and data export guarantees before contract signature; involve the legal team.
- Instrument all inference code with OpenTelemetry at the start of any observability program to preserve vendor-neutral optionality and minimize future migration costs.
- Evaluate the open-source ecosystem around any commercial platform: documentation quality, community size, and availability of third-party integrations.
- Establish a quarterly business review cadence with your primary observability vendor and document platform evolution commitments contractually where possible.

8.5 Takeaway

The AI observability vendor market offers genuine solutions at every tier, from open-source foundations to enterprise AI-native platforms to APM extensions. Still, no single vendor is the right choice for all organizations. The organizations that make good vendor decisions are those that understand their monitoring requirements before entering the market, apply a structured evaluation framework, and make architectural choices that preserve optionality rather than create lock-in at the instrumentation layer.

The vendor selection decision is not a one-time event. As the AI portfolio grows and as the vendor landscape continues to consolidate and expand, the platform strategy must be revisited. Organizations that treat their observability platform as a strategic investment, actively managing the vendor relationship and monitoring the market, will maintain the operational clarity their AI programs require.

9 Building Your Observability Architecture

A global insurance company runs forty-seven predictive models in production across underwriting, claims fraud, and customer churn. Each model was deployed by a different squad using a different tool: one team ships Prometheus metrics, another writes custom logs to an S3 bucket, and a third sends traces to a vendor-managed SaaS platform. When a senior regulator asks the VP of AI Operations for a single drift-status dashboard covering all models, the answer is three weeks of manual data wrangling, and a spreadsheet already stale by the time it's delivered. The architecture grew organically because no one designed it. Every squad solved its own visibility problem, and the organization now owns 47 siloed monitoring solutions that cannot communicate with one another. Designing the architecture before the sprawl sets in is not a technical preference; it is a management decision with direct consequences for audit readiness, incident response time, and the cost of every future model deployment.

Observability architecture determines whether your monitoring investment compounds or fragments. A poorly designed architecture forces every new model team to reinvent telemetry collection, storage, and visualization from scratch. Each reinvention introduces inconsistent schemas, incompatible retention policies, and bespoke alerting rules that no one outside the originating team can interpret. The result is an organization that spends more time maintaining and monitoring infrastructure than acting on the signals it produces. Senior managers who treat architecture as a purely technical concern discover too late that it governs the speed of every incident response, every audit response, and every board-level AI risk report they will ever produce.

The budget exposure is material. A federated architecture without a shared data layer results in duplicate storage costs, vendor licenses, and engineering labor. A centralized architecture without proper access controls exposes compliance risk when regulated data mixes with general telemetry. Getting the design wrong early means paying to redesign it under pressure, usually during an incident or a regulatory examination. The architecture decisions made in the first ninety days of an observability program set the cost structure and governance ceiling for years.

9.1 Architecture Patterns That Scale

Three architecture patterns dominate enterprise AI observability: centralized, federated, and hub-and-spoke. Each pattern reflects a different balance between operational control and team autonomy. Choosing among them is a management decision because the trade-offs map directly onto organizational structure, regulatory risk tolerance, and the speed at which the enterprise intends to scale its AI portfolio.

The Centralized Pattern

In a centralized architecture, a single observability platform receives telemetry from every model, pipeline, and inference endpoint in the organization. All metrics, logs, and traces flow to one collection layer, one storage backend, and one visualization tier. The ML platform engineering team owns and operates the entire stack. Model teams instrument their code according to a shared standard, typically OpenTelemetry, and push data to a central cluster of collectors. They consume dashboards and alerts that the platform team provisions.

The primary advantage is operational clarity. Drift metrics, latency histograms, and data quality scores for every model live in the same query interface. When a regulator requests a cross-portfolio drift report, the platform team generates it from a single warehouse query. The disadvantage is throughput: at enterprise scale, a single ingestion pipeline becomes a performance bottleneck and a single point of failure. Any outage in the central collector silences observability for the entire portfolio.

Centralized architectures work best for organizations with fewer than thirty models in production, a strong platform engineering team, and regulatory requirements demanding uniform data retention. Financial services firms subject to SR 11-7 often prefer centralization because it simplifies the evidence trail. The risk is that the platform team becomes a bottleneck for every new model deployment, slowing the pilot-to-production pathway if onboarding is not heavily automated.

Diagram 9.1 — Centralized Observability Data Flow

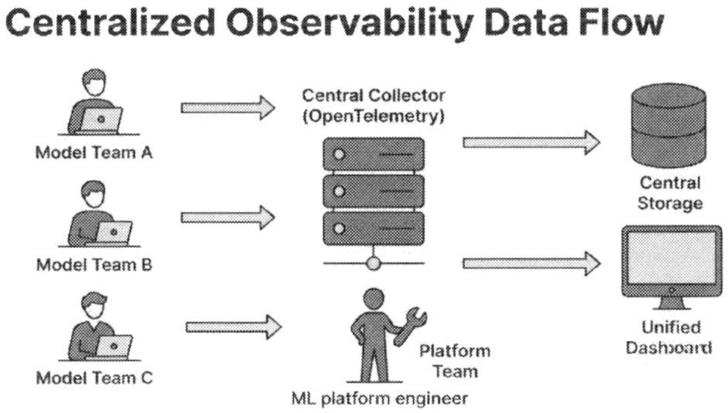

The Federated Pattern

A federated architecture distributes observability ownership to the teams that own each model or model domain. Each business unit operates its own collection, storage, and visualization stack. There is no central ingestion layer. Governance is enforced through policy rather than platform: the observability lead publishes schema standards, retention rules, and alert severity definitions, and each team implements them independently. Arize or Langfuse instances may be deployed per business unit, with Grafana dashboards maintained locally within each domain.

Federation is attractive because it eliminates the platform-team bottleneck. Model teams move faster when they control their own telemetry stack. They can tune retention, add custom metrics, and iterate on dashboards without waiting for a central team's sprint cycle. However, federation creates serious coordination costs at the executive level. Producing a portfolio-wide model health report requires aggregating data across incompatible schemas, different time-zone conventions, and varying retention windows.

The federated pattern suits organizations with high autonomy among business units, where data sovereignty rules prohibit centralization, or where the AI portfolio spans highly heterogeneous use cases. The observability lead in this model spends significant time enforcing schema compliance and building cross-domain aggregation pipelines. Without that investment, the federation drifts into unmanaged sprawl.

Diagram 9.2 — Federated Observability Domain Structure

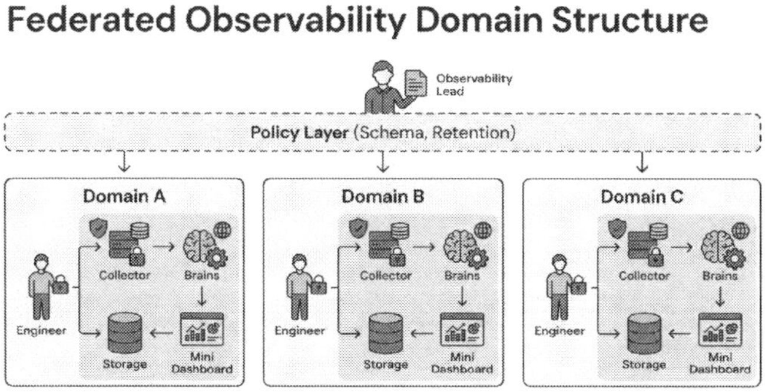

The Hub-and-Spoke Pattern

The hub-and-spoke pattern is the most common architecture in large enterprises because it balances central governance with local team autonomy. Each model domain operates a local spoke: its own OpenTelemetry collector, its own short-retention metrics store, and its own operational dashboards—the spokes forward a curated subset of telemetry to a central hub. The hub holds long-retention data, powers portfolio-level executive dashboards, and feeds the governance and audit reporting layer.

The key design decision is what travels to the hub and what stays in the spoke. Raw inference logs often remain local due to volume and cost; aggregated drift scores, SLA breach events, and model-level accuracy metrics are sent to the hub. OpenTelemetry's semantic conventions for generative AI give teams a shared vocabulary for LLM traces, reducing the translation cost between spokes.

Hub-and-spoke architectures require a governance document that defines precisely which signals must propagate to the hub, with what latency and granularity. Without that document, spoke teams make local decisions that silently degrade hub completeness. The observability lead must audit spoke-to-hub forwarding at least quarterly, comparing the hub's model inventory against the organization's model registry to detect models deployed but not reporting.

Diagram 9.3 — Hub-and-Spoke Telemetry Architecture

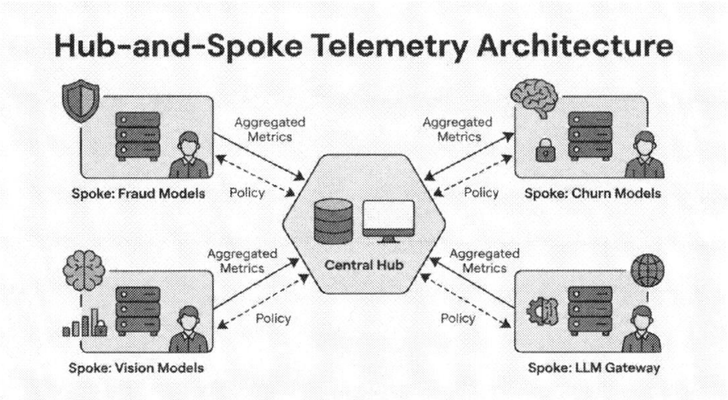

9.2 Data Flow, Storage, and the Observability Lake

Choosing an architecture pattern answers who owns the telemetry. Designing the data flow answers how that telemetry moves, where it lands, and how long it persists. Enterprise AI observability generates several distinct data types: real-time inference traces, batch drift statistics, data quality event records, model performance metrics, and audit-trail logs. Each type has different velocity, volume, and retention requirements. Forcing all of them through a single pipeline and a single storage layer produces either cost overruns or missing data at the worst possible moment.

Designing the Ingestion and Aggregation Pipeline

The ingestion layer is the first place where telemetry can be lost, corrupted, or delayed. For real-time inference telemetry, the ML platform engineer typically deploys an OpenTelemetry Collector as a sidecar or DaemonSet on the inference cluster. The collector receives spans from the model serving container, applies sampling and attribute filtering, and forwards events downstream. Sampling is a critical configuration choice: capturing every inference for a high-volume recommendation model generates terabytes of data per day, while oversampling a low-frequency fraud model may be necessary to detect rare failure patterns.

Aggregation sits between raw ingestion and long-term storage. Streaming aggregation systems compute rolling statistics such as prediction distribution histograms and population stability indices in near-real time. These aggregated statistics are far cheaper to store and query than raw inference logs, and they power operational dashboards and alerting rules. Batch aggregation runs on a scheduled basis to compute drift statistics that require a full population window, such as Kolmogorov-Smirnov test scores or Wasserstein distances. The pipeline design must account for both modes.

Schema management is the most frequently underestimated element of ingestion design. Every model team that connects a new spoke will instrument slightly differently unless the platform team enforces a schema contract at the collector level. OpenTelemetry's attribute validation and the use of a schema registry prevent schema drift before it reaches storage. A malformed event that bypasses the schema check and lands in the warehouse is far more expensive to remediate than one rejected at the collector boundary with a clear error message.

Diagram 9.4 — Ingestion Pipeline Stages

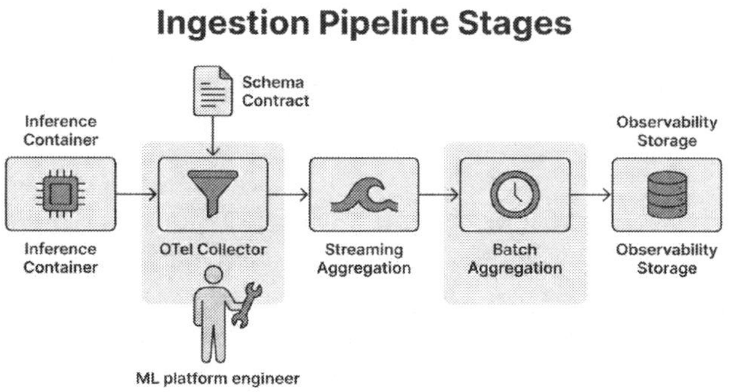

Long-Term Storage with Iceberg, Snowflake, and Databricks

Long-term observability storage is where management decisions about cost and governance intersect directly. Apache Iceberg has become the dominant open table format for observability data lakes because it supports time-travel queries and schema evolution without rewriting existing data. An ML platform engineer can add a new telemetry attribute to the inference trace schema without breaking queries that power existing dashboards, because Iceberg handles schema evolution at the metadata layer.

Snowflake and Databricks are the two most common compute layers on top of an Iceberg-formatted lake. Snowflake's separation of storage and compute allows the observability to run expensive cross-model drift queries on large clusters and then suspend them when finished, paying only for compute seconds consumed. Databricks Delta Live Tables automates the transformation pipeline that moves raw OpenTelemetry spans through aggregation stages to curated metric tables. Both platforms support role-based access control that maps cleanly to the observability RACI.

Retention policy design requires explicit decisions about which data classes belong in which tier. A practical three-tier model assigns real-time operational data to a hot tier with seven to thirty days of retention, aggregated drift statistics to a warm tier with one to two years of retention, and audit-trail logs to a cold tier with five to seven years of retention. These retention windows must appear in the observability governance charter and be reviewed annually because regulatory requirements such as SR 11-7 and the EU AI Act impose minimum retention periods that the data architecture must satisfy.

Diagram 9.5 — Three-Tier Observability Storage

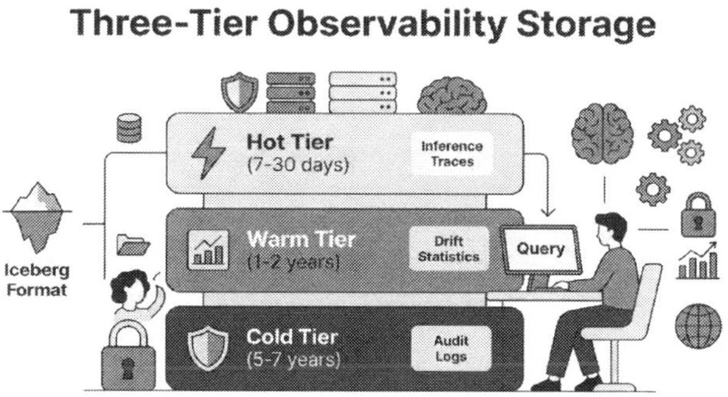

eBPF and Kernel-Level Telemetry

Extended Berkeley Packet Filter, universally abbreviated eBPF, represents a fundamentally different approach to telemetry collection. Traditional instrumentation requires developers to add observability code to application source files. eBPF programs run inside the Linux kernel, attaching to system calls, network events, and function entry points without any modification to the application being observed. For AI inference workloads, eBPF collectors capture GPU kernel execution times, memory copy operations between the CPU and GPU, and network throughput between inference nodes at sub-millisecond granularity, all without adding instrumentation to the model-serving code.

The operational significance for ML platform engineers is substantial. Many inference containers run third-party or compiled model-serving frameworks, making it impractical to insert OpenTelemetry instrumentation. eBPF sidesteps this problem by operating at the kernel layer below the application. Tools such as Pixie and Cilium Hubble use eBPF to provide automatic service-level metrics and network flow data for Kubernetes-hosted inference workloads. The performance overhead is typically under one percent of CPU, compared to three to five percent for user-space instrumentation in high-throughput serving environments.

eBPF telemetry complements rather than replaces application-level OpenTelemetry instrumentation. Kernel-level data captures infrastructure behavior: CPU scheduling, GPU memory pressure, and network retransmits. Application-level instrumentation captures semantic behavior: which prompt was processed, what the model returned, and how long post-processing took. The ML platform engineer configures the eBPF collector to tag kernel events with the same request identifier that the application writes into its OpenTelemetry spans, enabling queries that connect a user-visible latency spike to a specific GPU memory eviction at the kernel level during that same request window.

Diagram 9.6 — eBPF vs Application-Layer Telemetry

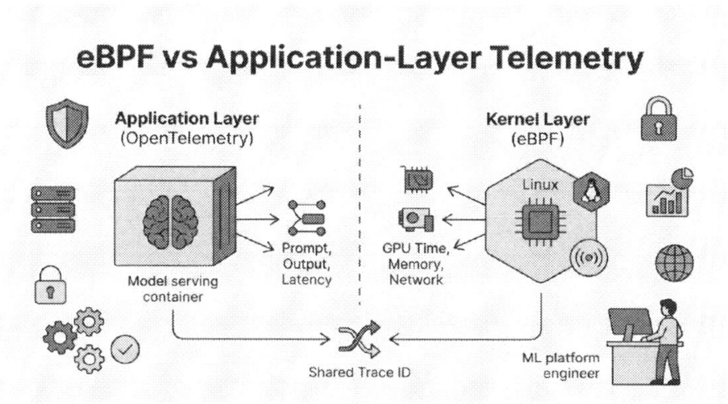

9.3 Dashboards, Integrations, and Reference Architecture

Telemetry that reaches storage but never reaches a decision-maker has no operational value. The visualization layer must maintain two distinct interfaces: the operational view that ML engineers and AI SREs use to detect and diagnose problems in real time, and the executive view that VPs and CIOs use to assess portfolio-level risk and compliance posture. Both views must draw from the same underlying data layer. Separate pipelines for each audience are a maintenance liability that typically diverges within six months of initial deployment.

Designing Dashboards for Two Audiences

The operational dashboard is a real-time instrument. It shows current inference latency percentiles, recent drift scores, data quality check pass rates for upstream feature pipelines, and active alert counts. An ML engineer or AI SRE using an operational dashboard needs to answer a specific question within seconds: is this model behaving as expected right now, and if not, which signal is most anomalous? Grafana is the most common platform for operational dashboards because its native integration with

Prometheus, OpenTelemetry, and Loki allows engineers to correlate metrics, traces, and logs in a single query session.

The executive dashboard is a portfolio risk instrument. A VP of AI Operations reviewing it needs to see how many models are within their performance SLA, the trend in data drift severity over the past 90 days, whether any model is approaching a retraining trigger, and whether any compliance-relevant alert was escalated to a human reviewer in the past week. These signals are computed from the warm-tier storage layer at a daily or weekly cadence and displayed as trend lines and status indicators.

The most common design error is building the executive dashboard first by pulling whatever is easiest to query and then retrofitting an operational view. The correct sequence is to define the operational signals first, because those are the raw measurements, and then define the aggregations that power the executive view. Every metric on the executive dashboard should have a traceable path back to a specific operational signal in the collection layer so that when an executive asks why a metric changed, the engineering team can answer with data.

Diagram 9.7 — Two-Audience Dashboard Design

Two-Audience Dashboard Design

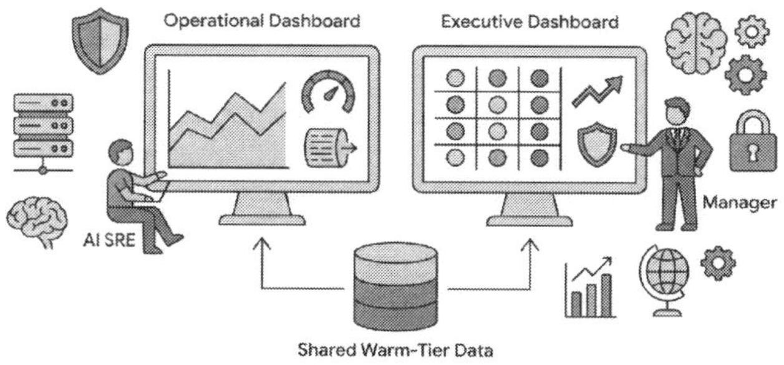

Integration Points Across the ML Lifecycle

An observability architecture that only monitors running models misses most signals that predict model failure. The most consequential integration points are upstream: the CI/CD pipeline that promotes models from staging to production, the model registry that tracks approved versions, the feature store that supplies input features at inference time, and the orchestration layer that schedules batch scoring jobs. When these systems emit events into the observability pipeline, the platform can correlate a production degradation with a specific upstream change.

CI/CD integration means that every model promotion event, including the model version, the evaluation metrics achieved in staging, and the name of the engineer who approved the deployment, is recorded as a structured event in the observability pipeline. This event serves as the baseline anchor: the system knows the model's expected performance characteristics at deployment, and drift is measured relative to that anchor. Model registries such as MLflow expose webhooks that the observability platform consumes to keep its model inventory synchronized with the organization's authoritative source of record.

Feature store integration is particularly important for detecting data drift before it reaches the model. When an upstream data pipeline changes, the feature store may begin serving features with a statistical distribution different from the one the model was trained on. If the observability platform monitors feature distribution statistics at the feature store's serving layer, it can detect this shift and fire an alert before the model's output distribution changes. This upstream detection shortens the time between a data quality event and the operational response by hours compared to monitoring only the model's output.

Diagram 9.8 — Observability Integration Across ML Lifecycle

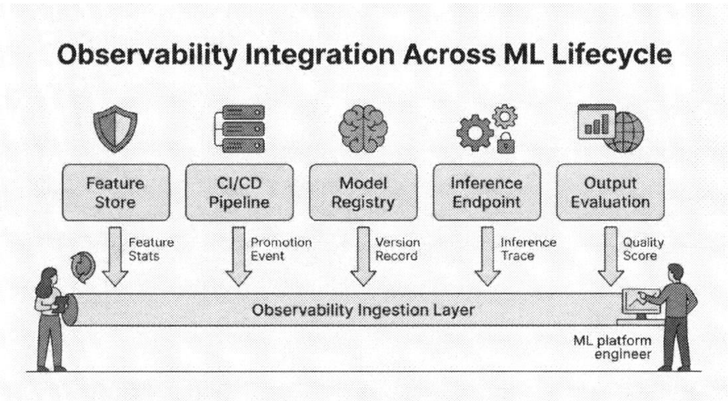

The Production-Ready Reference Architecture

A production-ready observability architecture combines the preceding elements into a coherent, governable stack. Instrumented model serving containers emit OpenTelemetry spans to a collector tier. eBPF agents on the same hosts capture kernel-level infrastructure metrics tagged with shared trace identifiers. The collector tier applies schema validation, sampling, and attribute redaction before forwarding events to the aggregation layer. Streaming aggregation computes real-time statistics for operational alerting; batch aggregation computes drift statistics for the drift detection layer. Both outputs land in an Iceberg-formatted data lake, partitioned by model identifier and date.

Above the storage layer, Databricks or Snowflake runs the transformation pipelines that produce curated metric tables, which are consumed by dashboards and alert evaluation engines. The hub-and-spoke control plane forwards portfolio-level aggregates to the central hub, where the observability lead and AI SRE team maintain the executive dashboard and cross-model alerting policies. The model registry and feature store emit lifecycle events directly into the ingestion pipeline, giving the architecture visibility into changes upstream of the model.

The governance layer sits above everything: access control policies that map organizational roles to data permissions, retention schedules that route data between storage tiers, and the audit-trail pipeline that writes every model performance event and alert escalation to the cold-tier archive. Without it, the observability stack is a technical capability with no connection to the organization's compliance posture. With it, the stack becomes the backbone of evidence for every AI governance review, every regulatory examination, and every board-level AI risk report the organization will produce.

Diagram 9.9 — End-to-End Reference Architecture Stack

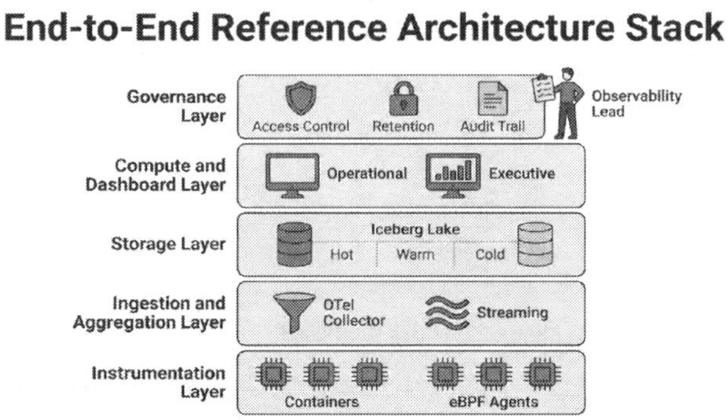

9.4 Manager's Checklist

- Select a hub-and-spoke or centralized architecture before the tenth model reaches production, and document the decision with rationale.
- Mandate OpenTelemetry instrumentation as a deployment gate: no model ships to production without an approved telemetry configuration.
- Define and publish a three-tier retention policy covering hot, warm, and cold storage with retention windows tied to regulatory requirements.
- Require the model registry and feature store to emit lifecycle events into the observability pipeline, so upstream changes are traceable.
- Audit spoke-to-hub forwarding completeness quarterly by comparing the hub's active model list against the model registry inventory.
- Establish a schema registry entry as a prerequisite for any new model domain connecting to the shared observability platform.
- Review the executive dashboard definition annually with the VP of AI and compliance leads to confirm metrics align with current regulatory expectations.

- Evaluate eBPF-based infrastructure telemetry for GPU-intensive inference workloads where user-space instrumentation adds unacceptable latency overhead.

9.5 Takeaway

Observability architecture determines whether every downstream investment in monitoring, alerting, and governance compounds or fragments. A deliberate choice among centralized, federated, and hub-and-spoke patterns, made before model sprawl sets in, reduces the long-term cost of observability by eliminating duplicate infrastructure and enabling portfolio-wide reporting from a single, well-governed data layer. The data flow design, storage tier structure, and integration points with the CI/CD pipeline and feature store are the structural elements that determine how quickly the organization can detect, diagnose, and respond to model failures in production.

The reference architecture described in this chapter is achievable with tools that most enterprises already license: OpenTelemetry for instrumentation, Iceberg for open table format storage, Snowflake or Databricks for compute, and Grafana or a commercial AI observability platform for visualization. What it requires is a management commitment to treat the architecture as a first-class deliverable, assign an observability lead with authority to enforce schema standards, and review the architecture's completeness annually as the AI portfolio grows.

10 Alerting, Incident Response, and Escalation

At 2:47 on a Tuesday afternoon, a PagerDuty alert fires, indicating that the fraud detection model serving a major retail bank's card-authorization pipeline has crossed its drift threshold. The alert goes to the on-call ML engineer, who acknowledges it and begins investigating. Four hours later, the engineer has not escalated; the model has been operating in a degraded state for the entire window, approving transactions at a rate fourteen percentage points above baseline—the fraud losses from that four-hour window total $340,000. The post-incident review reveals that the alert fired correctly, but the response playbook did not exist. The engineer did not know whether the remediation decision belonged to the ML platform team, the model owner, the fraud operations team, or the CISO. Technology worked. The process did not. Every serious AI observability program discovers that alert quality and response discipline are harder to build than the monitoring stack that generates the signals.

Alerting design is where observability converts from a data-collection exercise into organizational behavior. An alert that fires but produces no coordinated response is indistinguishable, from a business outcome perspective, from no alert at all. Senior managers who invest heavily in monitoring infrastructure and then leave response process design to individual engineers end up with a portfolio of alerts that engineers learn to ignore. Alert fatigue, the state in which a team receives so many notifications that it cannot distinguish urgent from routine, is not a technology problem; it is a governance failure.

The business stakes in poorly designed alert and incident response programs are direct. Revenue leakage occurs during every minute a degraded model remains in production without remediation. Regulatory exposure accumulates whenever a model failure affects a protected class or a regulated product and the organization cannot demonstrate a documented, timely response. Reputational damage follows when customers or counterparties experience the downstream consequences of an unresolved AI failure. Building the response infrastructure before incidents occur, including the RACI, playbooks, and integration with ITSM tools, is a management decision that separates organizations that recover quickly from those that absorb the full cost of every failure.

10.1 Designing Alerts That Drive Action

Alert design begins with a discipline that most engineering teams resist: deleting alerts rather than adding them. Every alert in a production system should be answerable with a specific action and owned by a named role. An alert that no one knows how to respond to, or that fires so frequently it becomes background noise, actively degrades the on-call team's ability to detect genuinely critical conditions. The observability lead must treat the alert catalog as a managed artifact, with a defined review cadence and a clear process for retiring alerts that have lost their operational value.

The Alert Fatigue Epidemic

Alert fatigue in AI monitoring programs follows a predictable pattern. Teams begin with wide-net alerting: every drift metric gets a threshold, every latency percentile gets a notification rule, and every data quality check fires an email on failure. Within 30 days of going live, on-call engineers receive dozens of notifications per shift. Most of these are informational, meaning no action is required. But because the alerts arrive through the same channel as genuinely critical notifications, engineers begin treating all of them with equal skepticism. When a model failure requiring immediate intervention occurs, it competes for attention against 40 routine drift nudges and 3 stale pipeline warnings.

Alert fatigue stems from not separating actionable signals from those that just need attention. An ML engineer monitoring a churn prediction model does not need a PagerDuty page at 3:00 AM because the model's precision dropped by 0.003 over the past week. That signal belongs in a weekly summary report consumed during business hours. A page at 3:00 AM is appropriate only when the model is causing real-time business harm, such as a pricing model generating quotes that violate rate-filing regulations.

Quantifying alert fatigue requires measuring the ratio of actionable alerts to total alerts over a rolling thirty-day window. An AI SRE team with below 30% actionable items is in the early stages of fatigue. Below 15%, the team has likely entered blanket-acknowledgment behavior, where alerts are acknowledged to silence notifications without investigation. The observability lead should report this ratio monthly, with a target of seventy percent or higher, before any new alert rule is added to the catalog.

Diagram 10.1 — Alert Fatigue Signal-to-Noise Ratio

Alert Fatigue Signal-to-Noise Ratio

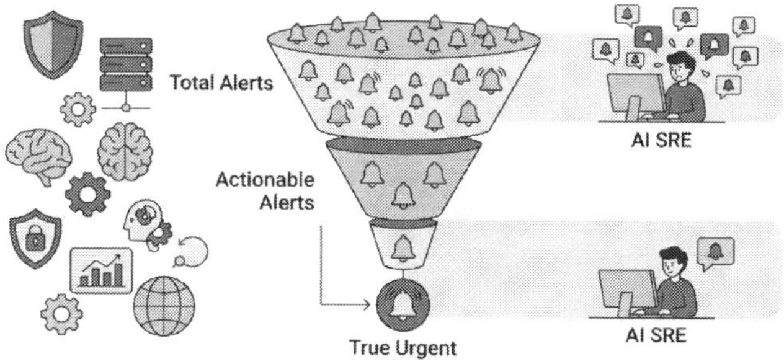

Building a Tiered Alerting Strategy

A tiered alerting strategy assigns each alert to one of four severity levels, each with a defined notification channel, response time expectation, and ownership. The informational tier captures slow-drift trends, minor data-quality anomalies, and performance metrics trending unfavorably but not yet crossing operational thresholds. These signals go into a dashboard digest or a scheduled Slack message. No human is woken up; the signal is consumed during the next business-hours review.

The warning tier captures conditions that are deteriorating and require investigation within four to eight business hours. An example is a feature store freshness delay that has exceeded the model's documented tolerance but has not yet caused measurable output degradation. A Jira ticket is created automatically; the model owner receives a notification during business hours and is expected to acknowledge and begin diagnosis within two hours. The critical tier captures conditions with confirmed or highly probable workflow-level impact: a model's primary business KPI has crossed a documented threshold, or a data pipeline has been failing for long enough that the model is operating on stale features.

The emergency tier is reserved for conditions requiring immediate human intervention, regardless of the hour. These are situations in which a production AI system is causing real-time harm: regulatory thresholds crossed, safety-critical outputs produced, or an SLA breach confirmed. Emergency alerts go directly to PagerDuty, with a 15-minute acknowledgment requirement and automatic escalation to the AI incident commander if the first responder does not respond. The distinction between critical and emergency is a management policy decision, not a technical one: it defines which failure conditions justify interrupting a human at any hour, and that decision must be made explicitly in advance.

Diagram 10.2 — Four-Tier Alert Severity Framework

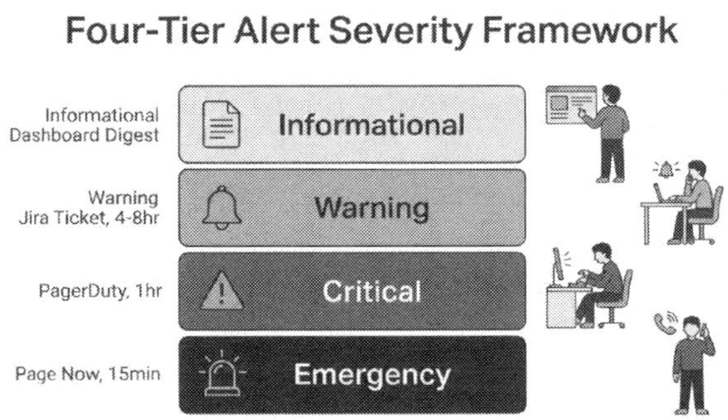

Integrating with ServiceNow, PagerDuty, and Jira

AI observability alerts must integrate with the ITSM ecosystem the enterprise already operates, not run parallel to it. Warning-tier alerts create Jira issues triaged during the next sprint review. Critical-tier alerts create ServiceNow incidents, triggering assignment-group routing, SLA timers, and management notifications. Emergency-tier alerts go directly to PagerDuty, reaching the on-call engineer's phone within seconds.

ServiceNow integration is achieved through its REST API or a pre-built observability vendor connector. The ML platform engineer configures the connector to populate the affected model identifier, alert severity, triggering metric and value, the threshold crossed, and a link to the diagnostic dashboard for that model. This structured population enables ServiceNow's routing rules to automatically assign the incident to the correct group, without requiring human interpretation of raw alert text.

PagerDuty schedules for AI incidents must account for the cross-functional nature of AI failures. A failure in a fraud model may require simultaneous involvement from the ML platform engineer, the model-owning data scientist, the fraud operations manager, and the compliance officer assessing regulatory exposure. PagerDuty escalation policies allow observability to configure multi-tier notification chains: the first page goes to the on-call ML engineer; if unacknowledged for 15 minutes, it escalates to the model owner; if unacknowledged for 30 minutes, it escalates to the AI incident commander, ensuring no incident stalls due to a single person's unavailability.

Diagram 10.3 — ITSM Integration by Alert Severity

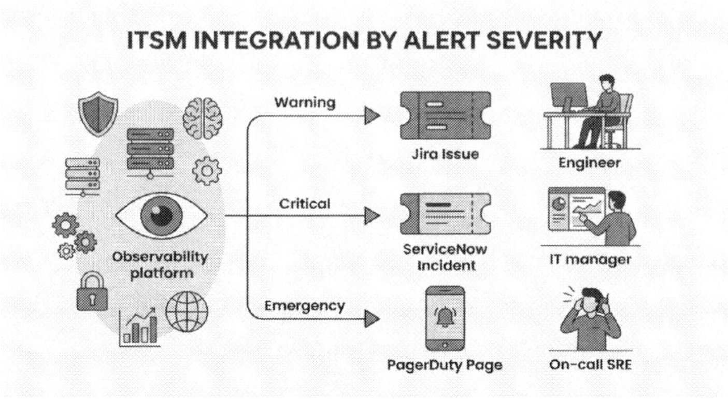

10.2 The AI Incident Response Playbook

A playbook is a pre-written response procedure for a specific class of failure. Without playbooks, every responder improvises under time pressure, producing longer resolution times, incorrect diagnoses, and incomplete documentation. For AI incidents, regulators examining a failure will ask for the documented response procedure that was followed, and its absence is itself a compliance finding. Playbook design is both an operational investment and a governance artifact.

The Five-Phase Incident Response Framework

Effective AI incident response follows five sequential phases: detect, diagnose, decide, deploy, and document. Detection is complete when the alert reaches the first responder and the responder confirms the alert represents a genuine production issue rather than a false positive or a test event. The first responder's initial confirmation action, typically a thirty-second review of the model's operational dashboard in Grafana or Arize, determines whether the incident is escalated or closed. The false-positive rate in the alert stream directly determines how efficiently this phase operates.

Diagnosis is the phase where the responder identifies the root cause. For a drift alert, diagnosis involves examining the feature distribution shift report, comparing the current input distribution against the training baseline, and checking the upstream data pipeline for recent changes. For a latency alert, diagnosis involves examining the infrastructure metrics, checking GPU utilization and memory pressure via the observability platform's infrastructure layer, and correlating the latency spike with any recent deployment events in the CI/CD pipeline. The playbook for each alert type should specify exactly which dashboard views to check and in what order, so that a responder unfamiliar with a specific model can diagnose efficiently.

Decision is the most consequential phase and the one most often left unplanned. Once the root cause is identified, someone must decide whether to roll back to a previous model version, route traffic to a fallback model, or continue operating in degraded mode while a longer-term fix is prepared. This decision requires authority beyond the on-call engineer. The playbook must specify who holds decision authority for each remediation class, aligned to the RACI. In the deploy phase, the chosen remediation is implemented and its effect verified in the observability dashboard. In the document phase, the full incident timeline is written into the ITSM record before the incident is closed.

Diagram 10.4 — AI Incident Response Five Phases

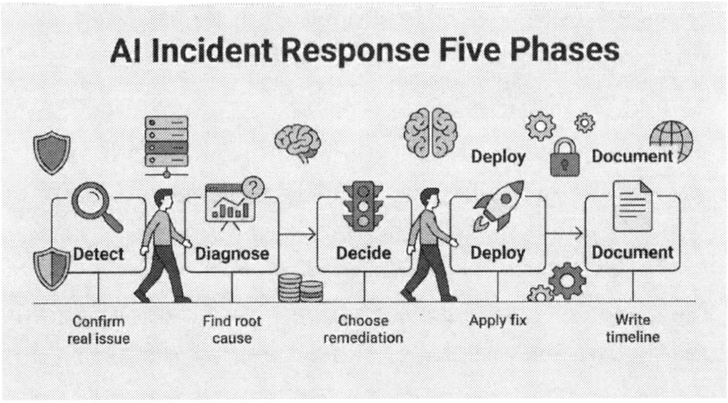

RACI for AI Incidents

A RACI matrix assigns each activity in the incident response process to one of four roles: Responsible, Accountable, Consulted, or Informed. For AI incidents, the RACI must address at least four dimensions of the response: technical investigation, remediation deployment, business impact assessment, and regulatory notification. Leaving any dimension unassigned produces a response gap that manifests as a delay at the most critical moment.

The ML platform engineer is typically Responsible for technical investigation and infrastructure-level remediation. The model owner, usually the data scientist who created and validated the model, is Accountable for the decision to roll back, retrain, or accept degraded operation. The AI SRE or observability lead is Responsible for monitoring remediation effectiveness and confirming the model has returned to acceptable operating parameters. The business stakeholder who owns the powered workflow is consulted on decisions to accept degraded operation, because that person understands downstream consequences better than the technical team.

Regulatory notification is the dimension of the AI incident RACI most frequently left unassigned. Certain failure classes trigger mandatory notification obligations: a credit-decisioning model producing disparate impact may require notification to the OCC; a clinical AI failure may require FDA notification. The RACI must identify who holds the Responsible designation for assessing regulatory notification obligations, and that assessment must occur within the first hour of the incident, not after it closes.

Diagram 10.5 — AI Incident RACI Matrix

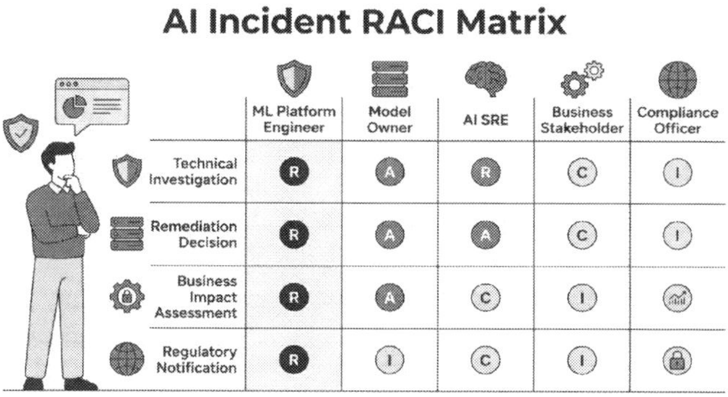

Automated Remediation and Human Judgment Boundaries

Automated remediation reduces the time from detection to resolution for a narrow yet important class of AI incidents. The key management decision is defining precisely which remediation actions are safe to automate and which require human authorization. This boundary is not a technical question; it is a risk-tolerance decision that must be made by the AI incident commander or the VP of AI Operations, in consultation with the legal and compliance team.

Safe automation candidates include routing inference traffic to a pre-validated fallback model when the primary model's error rate exceeds a configured threshold;

restarting a model serving pod that has entered an error state due to an infrastructure fault; and scaling up the inference fleet when latency percentiles cross a performance SLA threshold. These actions are reversible, they do not change the model's behavior, and they do not require any assessment of business or regulatory impact. Automated remediation for these cases can reduce the mean time to recovery from 20 to 30 minutes to under 2 minutes.

Actions that require human authorization include: rolling back to a previous model version, which changes the model's decision logic and may have regulatory implications; modifying the model's decision threshold or business rules, which changes its output distribution; and communicating to business stakeholders that the model is operating in degraded mode. These actions carry consequences that extend beyond the technical domain. The playbook must make the authorization boundary explicit: automated actions are executed by the observability platform's remediation engine, and human-authorized actions require a documented approval from the named Accountable role in the RACI before any change is made to the production system.

Diagram 10.6 — Automation Boundary in Incident Response

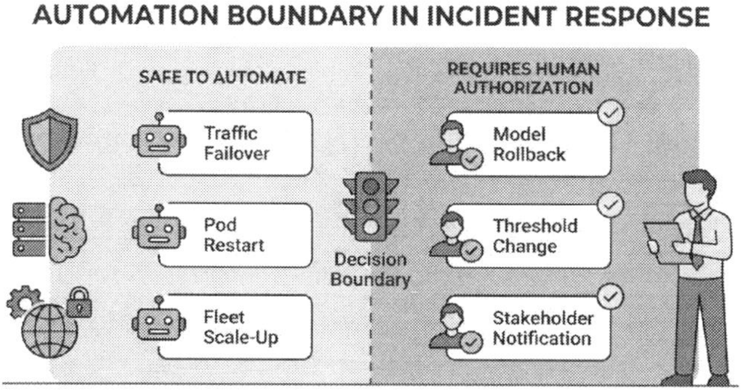

10.3 Post-Incident Review and Organizational Learning

The post-incident review is the mechanism by which individual AI failures become organizational capabilities. Without a structured review process, teams remediate incidents and move on, accumulating a body of unexamined experience that does not improve the organization's ability to prevent or respond to the next failure. With a structured review, each incident produces a permanent record of what failed, why it failed, what was done, and what systemic change will prevent recurrence. Over time, this record becomes the organization's most valuable guide to the actual failure modes of its AI systems in production.

Structuring the Post-Incident Review

A post-incident review for an AI failure should be completed within five business days of incident closure. The review document covers: a timeline from first detection to resolution, a root cause analysis identifying proximate and systemic causes, a quantified business impact assessment, a review of the response process against the playbook, and action items with owners and due dates.

The AI SRE or observability lead conducts the review with the model owner, the responding engineer, and affected business stakeholders.

Root cause analysis is the most technically demanding part of the review. Proximate causes in AI incidents often include data drift, upstream pipeline changes, or model degradation. Systemic contributing factors matter more for organizational learning: an absent playbook for this alert type, an uncalibrated threshold, a pipeline change not communicated to the model team, or an on-call rotation that lacked the right expertise. Each systemic factor produces an action item that improves the observability program's design.

The business impact assessment translates the incident's technical parameters into financial terms that senior management can act on. A drift incident that lasted four hours and affected a fraud model's false-negative rate by fourteen percentage points is more legible to a CEO as a $340,000 fraud-loss event than as a statistical deviation report. Quantifying impact in business terms also justifies observability investment: a year-end review showing that well-instrumented models had a 50% faster mean time to recovery than poorly instrumented ones is a compelling argument for increased program funding.

Diagram 10.7 — Post-Incident Review Document Structure

Post-Incident Review Document Structure

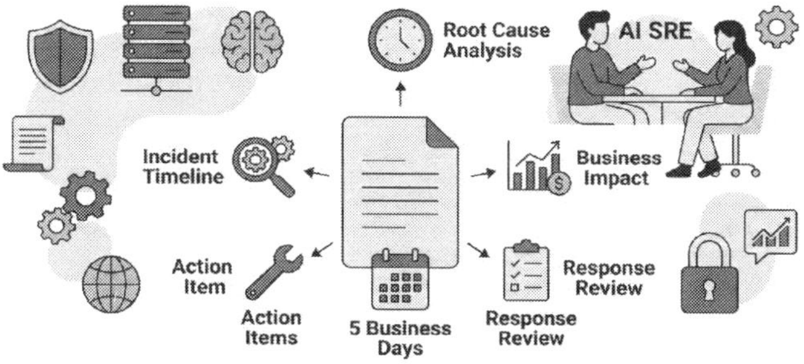

Building a Model Failure Knowledge Base

Individual post-incident reviews have limited value if stored as isolated documents. The organizational learning benefit comes from aggregating reviews into a searchable knowledge base that on-call engineers can consult during future incidents. When an ML engineer receives a drift alert at 11:00 PM for a model that has failed similarly twice before, the engineer should retrieve previous reviews and resolution paths within two minutes. That capability requires reviews to be stored in a structured, queryable format rather than in narrative documents on a shared drive.

ServiceNow's knowledge management module provides a natural home for the model failure knowledge base within organizations that already use ServiceNow for ITSM. Each post-incident review is stored as a knowledge article tagged with the affected model identifier, the alert type, the root cause category, and the resolution action taken. The observability platform or the ML platform team's Jira project board can link directly to relevant knowledge articles when a new incident matches a previous incident's pattern. Langfuse and similar LLM observability platforms are beginning to surface this capability natively for LLM incident patterns.

The knowledge base also drives the calibration of alert thresholds. If five reviews for the same model show the drift threshold fires at values that never caused measurable business impact, the observability lead has evidence to raise it. If three reviews show the latency alert fires only after a forty-five-minute performance window, the lead has evidence to lower the detection threshold. Alert threshold management should be a quarterly calibration activity driven by post-incident data rather than engineering intuition.

Diagram 10.8 — Incident Knowledge Base Feedback Loop

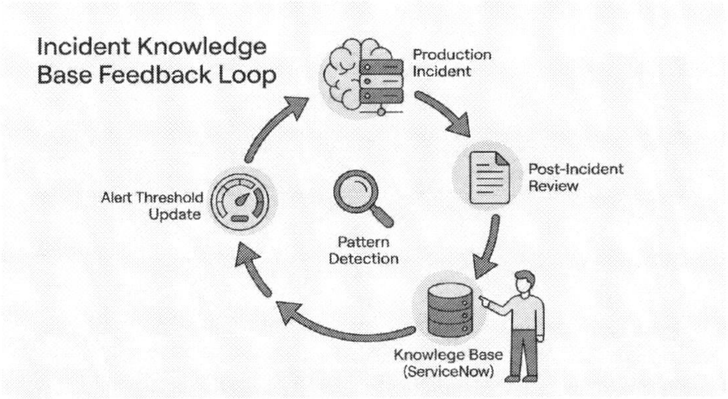

Measuring Incident Response Maturity

Incident response maturity is measurable through four metrics: mean time to detect, mean time to acknowledge, mean time to diagnose, and mean time to recover. Compute these per model, per domain, and across the portfolio, and report them monthly to the VP of AI Operations. Trends reveal whether investment in playbooks, RACI clarity, and ITSM integration is producing measurable improvement in response speed.

Mean time to detect measures the gap between when a failure first appears in telemetry and when the alert fires. A long value indicates that thresholds are too conservative, the monitoring cadence is too slow, or that certain failure modes are uncovered. Mean time to acknowledge measures response discipline: how quickly the on-call team acts. A long value indicates alert fatigue or on-call roster gaps that leave critical windows without adequate coverage.

Mean time to recover is the ultimate metric, most directly connected to business impact. Target recovery times vary by business criticality: a pricing model in high-frequency trading targets minutes; a weekly batch scoring model for customer segmentation targets hours. The observability lead should document target recovery times for every production model as part of its operational SLA and measure actuals against those targets in every post-incident review. This documentation creates the mission-aligned connection between technical response performance and business risk tolerance that senior management needs to make sound investment decisions.

Diagram 10.9 — Four Incident Response Metrics

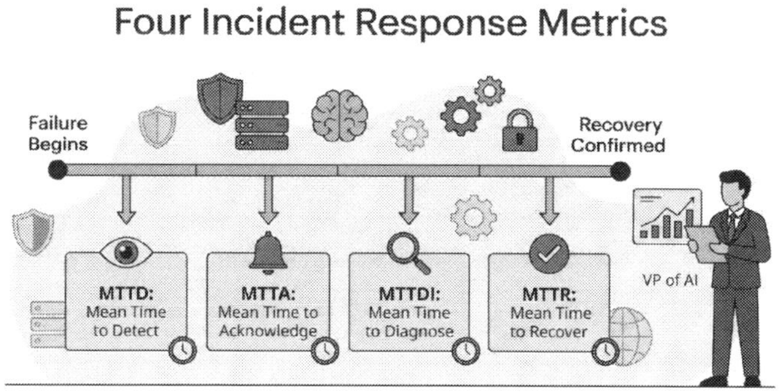

10.4 Manager's Checklist

- Define four alert severity tiers with specific notification channels, response time expectations, and ownership before the first production model is monitored.
- Measure the actionable alert rate monthly; target seventy percent or higher before adding any new alert rules to the catalog.
- Publish a written RACI for AI incidents covering technical investigation, remediation decision authority, business impact assessment, and regulatory notification.
- Configure PagerDuty escalation chains with at least three tiers so that no incident remains unacknowledged due to a single person being unavailable.
- Require a structured post-incident review within five business days for every critical or emergency AI incident, stored in a queryable knowledge base.
- Document a target mean time to recover for every production model as part of its operational SLA, and measure actuals monthly against those targets.
- Define the explicit boundary between automated remediation actions and actions requiring human authorization, and publish it as an organizational policy.
- Integrate AI observability alerts with ServiceNow and PagerDuty so that incident records are created, routed, and tracked through existing ITSM workflows.

10.5 Takeaway

Alerting and incident response are the operational layer that converts observability data into business protection. A tiered alert strategy, integrated with ServiceNow and PagerDuty, ensures that the right person receives the right signal at the right time, and that the response process is governed by an RACI that assigns accountability before the incident occurs. Alert fatigue is not a technology problem; it is a governance failure that requires the observability lead and VP of AI to jointly manage the alert catalog as a curated, maintained artifact rather than an accumulating list of thresholds.

Post-incident reviews are the investment that makes the entire incident response program improve over time. Each review produces specific improvements to thresholds, playbooks, and RACI assignments that reduce the duration and cost of the next incident. Organizations that conduct structured reviews and maintain a searchable knowledge base of AI failures build a compounding advantage in response capability. Those who treat each incident as a one-time event to be resolved and forgotten pay full price for every failure they encounter.

11 Governance, Compliance, and Audit Readiness

A regional bank deploys a consumer credit-scoring model in the first quarter. By the third quarter, the model had made credit decisions for 140,000 applicants. The bank's AI governance team has no structured observability program: model performance is reviewed quarterly by the owning data science team, but no drift logs, decision audit trail, or demographic parity statistics are retained beyond the immediate scoring run. When the OCC initiates an examination under SR 11-7, the examiner requests 18 months of model-monitoring records, a lineage map from training data to production output, and evidence that adverse action rates were reviewed for disparate impact at least quarterly. The bank cannot produce any of these records. The resulting matter-requiring-attention citation costs eight months of remediation work, halts new model deployments, and draws board-level scrutiny the AI team had hoped to avoid. Properly designed observability would have produced all of these artifacts automatically. The examination finding is not a technology failure; it is a program design failure.

AI governance is not a compliance checkbox that sits beside the technical work. It is the organizational process that converts observability data into defensible evidence of responsible operation. Regulators, auditors, and board risk committees do not ask whether an AI system is technically sophisticated; they ask whether the organization can demonstrate that it knows what the system is doing, that it has controls over that behavior, and that it responds appropriately when the system deviates from its intended purpose. Observability is the mechanism that makes those demonstrations possible.

The regulatory landscape for enterprise AI is expanding rapidly. Banking institutions must comply with SR 11-7; federal agencies must comply with OMB M-26-04; healthcare organizations must comply with HIPAA and FDA guidance on software as a medical device; any organization operating in the European Union must address the EU AI Act's conformity assessment and logging requirements. The NIST AI Risk Management Framework provides a voluntary but increasingly expected cross-industry baseline. A GRC lead who has not mapped the AI portfolio to applicable regulatory requirements and verified that the observability program produces the required evidence has left the organization exposed.

11.1 The Regulatory Landscape for AI in Production

No single regulatory framework governs AI across all industries and geographies. Senior managers must understand which frameworks apply to their portfolio, what evidence those frameworks require, and how the observability program produces that evidence. Treating this as a legal team problem rather than a program design problem produces the scenario at the start of this chapter. The GRC lead and the ML platform lead must design the observability stack jointly, with regulatory evidence requirements as first-class design inputs.

SR 11-7, NIST AI RMF, and the EU AI Act

SR 11-7, the Federal Reserve's 2011 supervisory guidance on model risk management, remains the most operationally detailed regulatory framework for AI governance in the United States financial services sector.

It requires financial institutions to document model development, validation, and ongoing monitoring with sufficient rigor that a knowledgeable independent reviewer can evaluate the model's conceptual soundness. SR 11-7 demands evidence of ongoing performance monitoring against established benchmarks, periodic review of model assumptions and data inputs, and documentation of any model use or limitations differing from the original validation scope.

The NIST AI Risk Management Framework, published in January 2023, provides a voluntary but increasingly referenced cross-sector governance structure organized around four functions: Govern, Map, Measure, and Manage. The Measure function is most directly relevant for observability: AI risks should be evaluated using quantitative and qualitative methods, measurement results should feed the management function, and measurement activity should be documented with sufficient specificity to support external review. Many federal procurement processes now require NIST AI RMF alignment as a condition of contract award.

The EU AI Act, which entered into force in August 2024, creates the most prescriptive observability requirements yet seen in AI regulation. High-risk systems, including those used in credit scoring, employment decisions, and biometric identification, must maintain technical documentation, keep operation records sufficient to enable post-hoc analysis, and log automatically generated decisions with enough metadata to identify the system version and data inputs used. The logging layer must capture decision records in a form that reconstructs the context of any individual automated decision, and retain them for the period specified by the Act.

Diagram 11.1 — Regulatory Framework Coverage Map

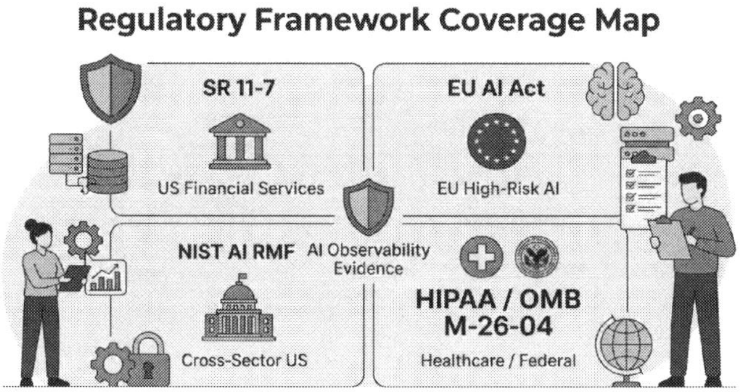

HIPAA and OMB M-26-04 Requirements

Healthcare organizations deploying AI that processes protected health information face a dual compliance burden. HIPAA's Security Rule requires administrative, physical, and technical safeguards for systems that process PHI, including AI systems that ingest clinical data. The observability program must ensure that monitoring systems handling PHI are covered under the Business Associate Agreement framework and that access to PHI-containing inference logs is restricted to authorized personnel. An observability platform that aggregates inference traces with patient identifiers without proper access controls creates a HIPAA violation, independent of the underlying model's clinical performance.

OMB Memorandum M-26-04, issued in April 2025, establishes requirements for federal agencies deploying AI in high-impact functions. It requires agencies to maintain an AI use-case inventory, conduct annual risk assessments for high-impact systems, and ensure that monitoring and logging capabilities support human oversight. The annual risk assessment requirement aligns with the observability program's model performance review cadence; the logging requirement aligns with the audit-trail architecture layer.

Across all frameworks, a consistent pattern emerges: regulators want evidence that a responsible team was monitoring the AI system's behavior, that anomalies were detected and addressed, and that activity records are available for examination. The observability program that produces structured, queryable records of model performance, drift events, alert responses, and remediation actions is, in essence, the governance program regulators are asking for. The gap between having a platform and having a defensible program is the gap between collecting data and retaining it, structuring it, and making it accessible in the form auditors require.

Diagram 11.2 — HIPAA and OMB Observability Requirements

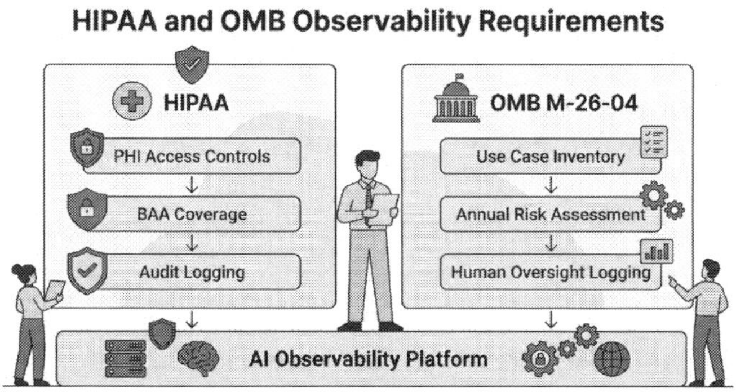

The Three Cs of AI Governance

Control, Clarity, and Continuity form the practical governance triad that transforms regulatory compliance from a reactive exercise into a sustainable operational capability. Control means that the organization can demonstrate defined boundaries on AI system behavior, enforced through technical mechanisms rather than policy alone, with deviations detected and addressed. From an observability perspective, control is demonstrated by alert policies that fire when models exceed approved thresholds, remediation workflows that require documented human review before a threshold-crossing model continues, and access controls that prevent unauthorized configuration changes.

Clarity means the organization can explain its AI systems in terms that non-technical stakeholders, including regulators and auditors, can understand. A model card documents the model's intended use, training data sources, performance metrics across demographic subgroups, known limitations, and the operational parameters under which it was validated. Observability data populates the card's ongoing performance section: each quarterly review updates it with current metrics, drift statistics, and any changes to the model's operating environment.

Continuity means the governance program survives personnel changes and organizational restructuring. A program depending on one person's knowledge of where monitoring data lives is a personal service that terminates when the person leaves. Continuity requires documented governance processes, observability data stored in records-managed systems, and audit-evidence retrieval procedures tested at least annually. The observability lead should conduct an annual continuity drill to assess whether an incoming lead can produce a complete regulatory examination package using only documented procedures and existing platform data.

Diagram 11.3 — The Three Cs of AI Governance

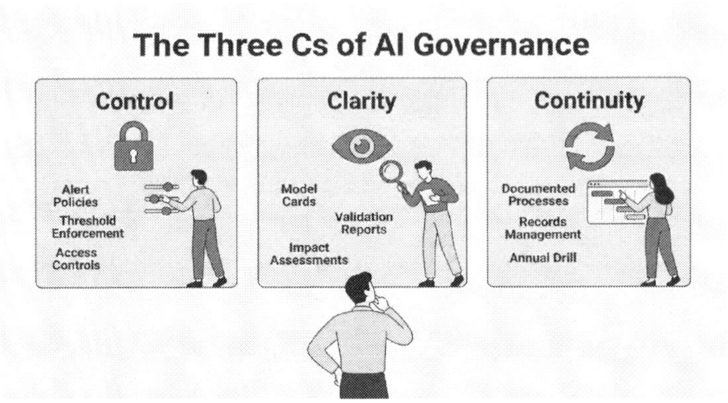

11.2 Observability as Evidence and the Audit Trail

The difference between an observability program and a defensible governance program is not the quality of the dashboards; it is the quality of the records. Dashboards support real-time decision-making. Audit evidence is a structured, immutable, attributable record of what the AI system did, what the monitoring program detected, how the team responded, and what changed as a result. Building the audit evidence layer requires deliberate choices about what data is retained, in what format, with what access controls, and for how long.

Turning Observability Data into Audit Artifacts

Not all observability data is audit evidence. Raw inference logs are operational data; most individual records have no governance significance. Audit evidence consists of structured records demonstrating governance activity: model performance reviews with model owner sign-off, drift alerts with documented response actions, threshold change requests with approval records, and monitoring reports delivered to named reviewers at specified dates. The observability platform must produce these structured records as first-class outputs, not as afterthoughts reconstructed from raw logs when an examination arrives.

The audit trail for a production AI model should document five event categories: model lifecycle events such as deployment, version updates, and retirement; performance monitoring events such as scheduled metric reports and drift alert firings; response events such as incident acknowledgements and post-incident reviews; governance review events such as quarterly and annual validation reviews; and data lineage events such as feature schema changes and training data updates. Each record should include event type, timestamp, model identifier, the acting role or system, and a structured payload sufficient to reconstruct what happened.

Immutability is a non-negotiable characteristic of audit evidence. A monitoring record that can be modified after the fact is not reliable evidence. The cold-tier storage layer should be configured with write-once semantics: records can be written but not modified or deleted during the retention period. Apache Iceberg supports immutable snapshot retention; Amazon S3 Object Lock and Azure Blob Storage provide the underlying mechanisms. The GRC lead should verify the cold-tier configuration is active at least annually and has not been modified by an infrastructure change.

Diagram 11.4 — Five Categories of AI Audit Evidence

Five Categories of AI Audit Evidence

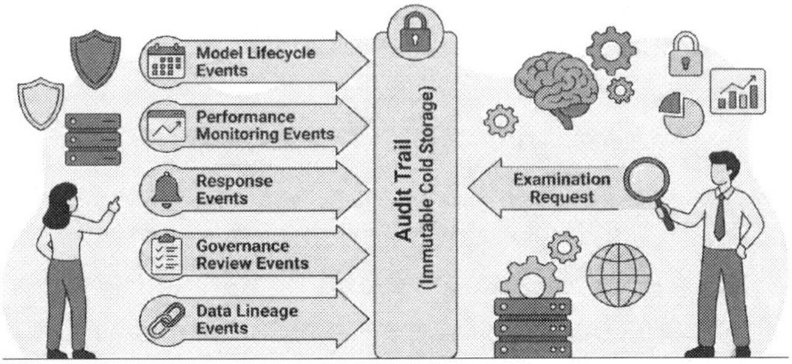

Model Cards, Validation Reports, and Lineage

A model card is the primary documentation artifact for an AI system's governance record. Adopted by MLflow, Hugging Face, and most enterprise model registry platforms, the model card records the model's intended use, training data characteristics, performance across relevant subgroups, known limitations, and validation conditions. In a governance context, the model card is not a one-time deployment document; it is a living record updated at each performance review. The observability lead ensures each model's card reflects current monitoring data.

Validation reports document the independent review that assessed the model's fitness for its intended use before deployment. SR 11-7 requires validation by personnel independent of model development; the EU AI Act requires evidence of conformity assessment. The validation report is stored in the model registry as an immutable artifact linked to the specific model version. When observability data indicates production behavior has diverged significantly from validated behavior, a re-validation event should be triggered and documented before any decision to continue operating the model.

Data lineage traces a model's output provenance back through inference-time feature values, feature transformation logic, raw data sources, and the training dataset. Full lineage is essential for investigations of individual decisions, for regulatory examinations that ask how a specific adverse action was produced, and for root cause analysis of failures originating from data quality issues. OpenLineage and Marquez provide open-source lineage tracking for pipelines; Databricks Unity Catalog and Snowflake's lineage features extend this to compute and transformation layers. Lineage data must be captured at each stage and stored in a queryable format to support specific decision inquiries.

Diagram 11.5 — Model Card, Validation, and Lineage Chain

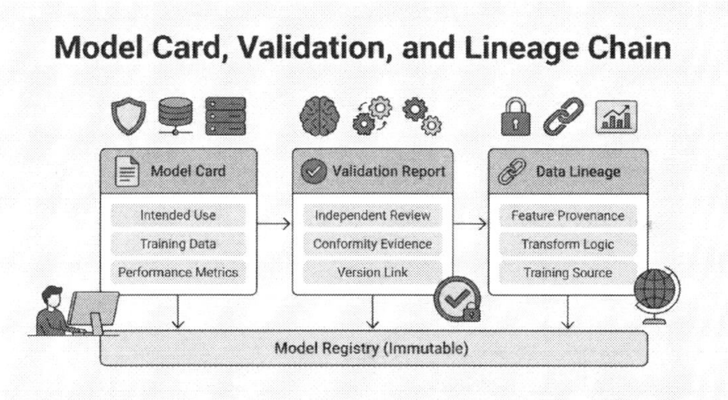

Access Controls and Segregation of Duties

Access control for AI observability is a governance requirement that most organizations address late. The principle is straightforward: the engineers who build and operate models should not have unilateral control over the monitoring data used to assess those models. An ML platform engineer who can modify drift thresholds, suppress alerts, and delete monitoring records without a second approver represents a control failure regardless of integrity, because the control framework depends on process, not trust. Segregation of duties assigns write access to monitoring configurations to the platform team and read access to audit records to an independent GRC function.

Role-based access control should map to the AI RACI. Model owners can read their model's performance data and submit threshold-change requests but cannot approve them. The observability lead can modify operational alert thresholds within documented policy limits but cannot delete or modify cold-storage audit records. Only a named infrastructure administrator, whose actions are logged and reviewed quarterly, can perform storage-level operations on the cold tier. Controls are documented in the observability security policy and verified through an annual access review.

Third-party observability vendors, including Arize, Langfuse, and commercial APM platforms, introduce supply chain access considerations. When a vendor platform processes inference traces, the vendor's engineers may have access to that data under the service agreement. The GRC lead must review vendor access controls, data processing agreements, and subprocessor lists during vendor assessment. For HIPAA-covered models, the vendor relationship requires a Business Associate Agreement. For high-risk systems under the EU AI Act, the data processing arrangement must comply with GDPR transfer requirements.

Diagram 11.6 — Segregation of Duties in Observability

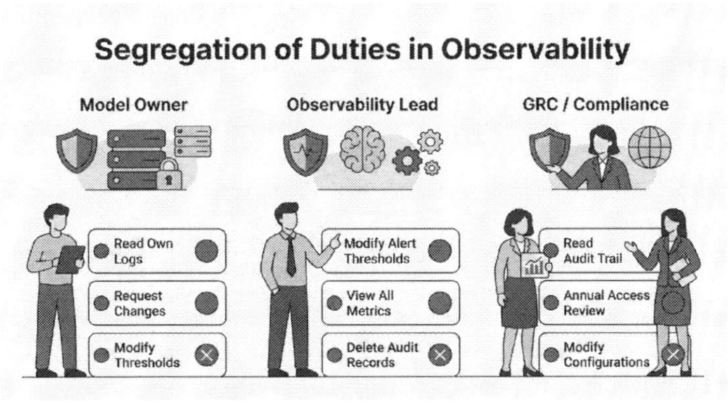

11.3 Building the Audit-Ready Governance Program

An audit-ready AI governance program operates continuously so that every examination finds the program already functioning, not assembled under pressure. The difference between organizations that pass AI governance examinations with minimal findings and those that receive material remediation orders is rarely the quality of their models. It is the quality of their records, the clarity of their process documentation, and the speed with which they retrieve specific evidence in response to specific questions.

Connecting Observability to the GRC Program

The GRC program tracks regulatory requirements, assesses compliance status, and manages remediation of identified gaps. Most enterprises run their GRC program on ServiceNow GRC, Archer, or OneTrust. Connecting the AI observability program to the GRC platform means that model-monitoring findings, drift alerts, and post-incident reviews flow automatically into the GRC evidence repository, tagged with the specific regulatory requirements they satisfy. This eliminates the manual evidence-gathering that consumes weeks of engineering time when observability and GRC programs operate in isolation.

The practical integration point is the governance review event. When the observability lead conducts a quarterly model review, the review artifact, which includes current performance metrics, drift statistics, alert activity, and remediation actions, is automatically published to the GRC platform as an evidence record. The GRC lead can view evidence library completeness at any time, see which models have current review records and which are overdue, and produce a regulatory readiness report without contacting the ML platform team.

This integration also supports proactive gap identification. When a new regulatory requirement is published, the GRC maps it to the required evidence types. The system can then identify which models lack the required evidence and create remediation tasks for the observability lead. This responsible-by-design approach, where governance requirements drive observability configuration rather than the reverse, is the operational model that keeps mature AI governance programs ahead of the regulatory cycle.

Diagram 11.7 — Observability to GRC Evidence Flow

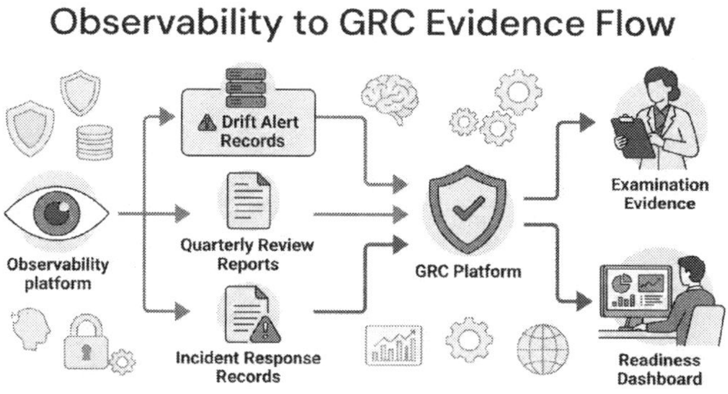

What Regulators Actually Ask For

Enterprise teams that have never experienced an AI regulatory examination tend to prepare for the wrong questions. Examiners are not interested in architectural sophistication or backtesting accuracy metrics. They ask operational questions: Who is responsible for monitoring this model? When was it last reviewed? What changed as a result? Can you show the monitoring record for the period when this adverse event occurred? These questions require the ability to produce structured records on demand, not to explain the algorithmic approach.

A practical preparation exercise is to run a mock examination using standard examiner questions: produce the model monitoring report for the third quarter; show the alert history for the past six months; provide evidence that this model was reviewed for demographic parity in the past twelve months; show the training data lineage for this model version. Each question maps to a specific data requirement, traceable to an element of the observability architecture.

Preparation also requires understanding the regulator's institutional context. SR 11-7 examiners focus on whether model risk is actively managed and whether the bank can identify failures before they become safety-and-soundness issues. EU AI Act assessors focus on the completeness of documentation and the traceability of automated decisions. HIPAA auditors focus on access controls and the use of minimum-necessary PHI. Tailoring the evidence package to the specific regulator's concerns demonstrates the organizational maturity examiners are evaluating.

Diagram 11.8 — Examiner Questions Mapped to Evidence

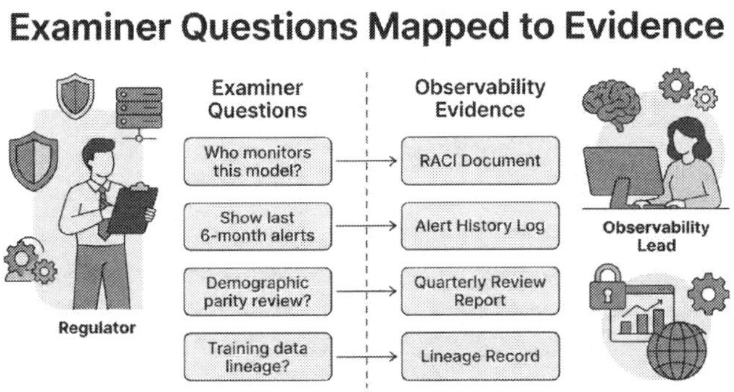

Building Governance Cadence Into Operations

Governance that runs only when an examination is anticipated does not actually govern. A mature AI governance program runs on a fixed calendar: weekly operational reviews for high-risk models, monthly health reports for all production models, quarterly performance and drift reviews with model-owner sign-off, and annual full validation reviews for models in scope for SR 11-7 or the EU AI Act high-risk classification. Each cadence produces a specific artifact stored in the GRC evidence repository and available for examination at any time.

The quarterly performance review is the governance activity most directly supported by the observability program. The observability lead pulls a structured report for each model that shows: performance metrics for the quarter, drift alerts and their resolutions, operating environment changes, and current performance versus the validated baseline. The model owner signs off with a documented attestation that the model remains fit for its intended use, or that a remediation plan is in place. This sign-off demonstrates active human oversight, the core expectation of every major AI regulatory framework.

Embedding governance cadence into the observability platform's workflow automation makes it sustainable at scale. When a quarterly review date approaches, the platform automatically generates the draft review package. It sends it to the model owner via a ServiceNow task or Jira ticket with a sign-off deadline. If sign-off does not occur, an escalation goes to the VP of AI Operations, and the item is logged as open in the GRC platform. This automation converts governance from an activity dependent on individuals remembering it into a process that the organization enforces through its existing workflow infrastructure.

Diagram 11.9 — AI Governance Calendar Cadence

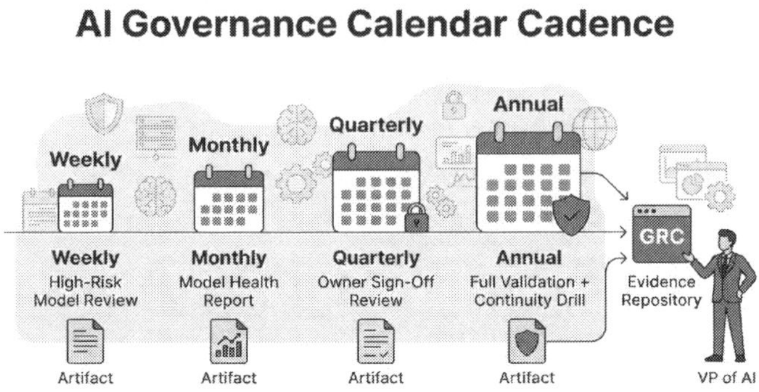

11.4 Manager's Checklist

- Map every production AI model to the applicable regulatory frameworks and document the mapping in the GRC platform.
- Require a model card for every production model, updated at each quarterly review with current observability-sourced performance data.
- Configure cold-tier storage with write-once semantics for audit-trail records and verify the configuration annually.
- Integrate the observability platform's quarterly review workflow with the GRC platform so review artifacts are automatically tagged to regulatory requirements.
- Publish a documented segregation of duties policy for observability access controls and conduct an annual access review against it.
- Require a data lineage record from training data to production output for every model subject to SR 11-7 or EU AI Act high-risk classification.
- Run a mock regulatory examination at least annually using standard examiner questions to identify evidence gaps before a real examination.

- Establish a quarterly model performance review cadence with documented model owner sign-off, escalating overdue reviews to the VP of AI Operations.

11.5 Takeaway

Observability is the technical backbone of AI governance, but governance requires more than a monitoring platform. It requires structured records retained with immutability controls, accessed through documented role-based policies, and organized so that specific regulatory questions can be answered with specific evidence in a defined response time. The organizations that navigate AI regulatory examinations with minimal findings are those that designed their observability programs with governance outputs as first-class requirements.

The governance cadences described in this chapter, from weekly operational reviews to annual continuity drills, convert technical capability into demonstrable accountability. A mature governance program does not fear regulatory examination; it demonstrates the operational clarity and responsible-by-design discipline that the observability program makes possible.

12 Cost Optimization and FinOps for AI

A technology company's VP of Engineering receives a cloud bill showing that AI-related infrastructure costs increased by 340% over 12 months. The organization deployed fifteen new AI features across three product lines during that period. Still, the breakdown is troubling: sixty-two percent of the cost is attributable to inference serving, and the team cannot identify which features, user segments, or model versions are driving that spend. The LLM gateway does not log token counts per feature. The GPU cluster runs inference jobs from twelve model teams on shared infrastructure with no cost tagging. The storage layer retains raw inference logs indefinitely. Three months after the VP requests a cost attribution report, the FinOps lead produces a rough estimate rather than an auditable number. The organization is spending $4.2 million per year on AI infrastructure and cannot explain more than thirty percent of it at the feature level. Without observability data that captures cost signals alongside operational signals, FinOps for AI is guesswork.

AI infrastructure costs are qualitatively different from traditional application infrastructure costs because the unit economics are opaque by default. A traditional web application's cost scales predictably with the number of requests and compute time. An AI application's cost scales with token consumption, model size, inference batch configuration, context window length, and GPU memory allocation, none of which appear in a standard cloud bill without explicit instrumentation. A senior manager who approves an AI budget without understanding these cost drivers is approving a number that will change unpredictably as usage patterns evolve.

The FinOps discipline requires adaptation for AI workloads because the standard FinOps toolkit was designed for compute and storage, not for token consumption and model inference. The AI-specific cost management capability described in this chapter integrates with the observability program to capture cost signals alongside operational signals, enabling cost attribution at the model, feature, and user-segment level. This integration converts AI infrastructure spending from unallocated overhead into manageable, accountable cost centers whose unit economics can be optimized, negotiated, and forecast.

12.1 Understanding the AI Cost Structure

Before an organization can control AI costs, it must understand them. The AI cost iceberg presents a deceptively small visible surface above a much larger submerged structure of token costs, storage costs, data transfer costs, and engineering labor. A FinOps lead who maps only the visible surface will undercount the total AI cost of ownership by fifty to seventy percent and will design optimization initiatives that target the wrong components.

The AI Cost Iceberg

The visible components of AI infrastructure cost are compute and API fees: GPU instance hours for model serving, API call fees for third-party LLM services, and batch inference compute for scheduled scoring jobs. These appear in the cloud provider's billing console and are what finance teams typically track as the AI budget. They are also the easiest to reduce through right-sizing and scheduling optimization, which is why most AI cost programs focus almost exclusively on them.

The submerged components are larger and often grow faster. Storage costs accumulate as inference log retention policies are left at default settings. Data transfer costs arise when inference nodes pull features from stores in different availability zones. Model evaluation costs compound as LLM-as-a-judge pipelines run evaluation calls on every production inference, effectively doubling token costs. Engineering labor to maintain serving infrastructure, run retraining pipelines, and manage vendor relationships is the largest submerged component and the one least frequently included in AI budget discussions.

Quantifying the total AI cost of ownership requires a cost inventory covering all components for each model or AI feature. The observability program is the primary data source because it captures the signals that map to cost drivers: token counts per inference, GPU utilization per serving pod, feature store query counts per model, and storage bytes per data class. Without this data, the cost of inventory is based on estimates and list prices rather than actual consumption, resulting in attribution that is inaccurate enough to mislead budget decisions.

Diagram 12.1 — The AI Cost Iceberg

The AI Cost Iceberg

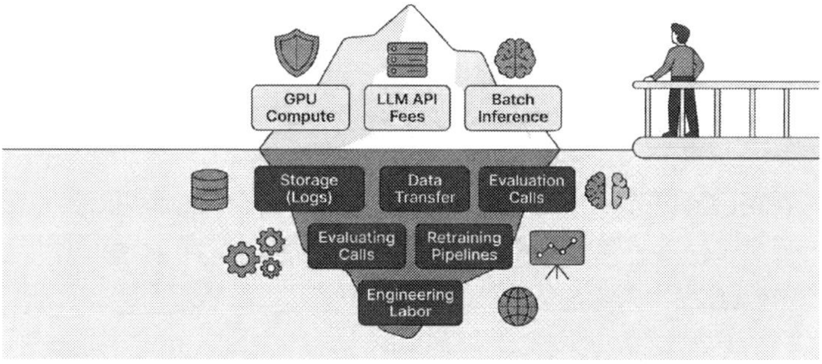

Token Economics for LLM Workloads

Token economics is the discipline of understanding, attributing, and optimizing the token consumption that drives LLM API costs. Input and output tokens are priced separately, with output tokens typically costing two to four times as much as input tokens for the same model. An organization that deploys an LLM-powered feature without understanding its token consumption profile has no basis for forecasting API costs as the feature scales, and no framework for evaluating whether a cheaper model would serve the use case at a lower cost.

Token tracking requires explicit instrumentation in the LLM gateway or application layer. The observability platform should capture, for every LLM inference call: the model identifier, prompt token count, completion token count, requesting feature identifier, user segment identifier if available, and timestamp. Langfuse and LangSmith provide native token tracking; OpenTelemetry's GenAI semantic conventions standardize attribute names, so token data from different teams use a consistent schema in the shared observability lake.

The FinOps lead uses token consumption data to compute cost per unit of business value for each LLM feature: cost per customer service ticket resolved, cost per document summarized, cost per user query answered. These unit economics make each feature's cost visible and comparable. When unit economics are unfavorable, the standard levers are: prompt compression to reduce input token count, model downgrade to a smaller model for simpler queries, response caching for frequent identical prompts, and context window management to avoid unnecessarily long conversations.

Diagram 12.2 — Token Economics Attribution Flow

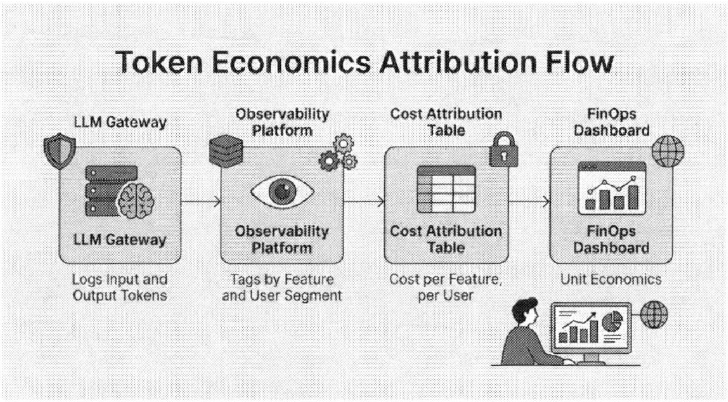

Cost Attribution Models for Enterprise AI

Cost attribution assigns AI infrastructure costs to the business units, products, and features that generate them. Without attribution, AI costs appear as undifferentiated overhead on the engineering cost center, giving no business unit a financial incentive to optimize its AI consumption and giving no executive a view of whether a specific investment is generating sufficient return. Attribution converts AI infrastructure spending into manageable, accountable cost centers.

The most granular attribution model allocates costs at the model-call level: each inference call is tagged with the feature identifier, business unit identifier, and user segment identifier, and the cost is computed from the token count and the applicable rate card. A simpler model allocates costs at the model level: the total monthly cost of each deployed model is divided among consuming business units in proportion to their share of inference volume. The simpler model is less precise but implementable without changes at the application layer, using existing observability data.

Chargeback transfers attributed AI costs to the consuming business unit's budget, creating a direct financial consequence for consumption decisions. Showback reports attributed costs to business unit leaders for visibility without transferring budget. Most organizations start with showback because chargeback requires finance system integration that takes time to establish. The FinOps lead should design the attribution model to support chargeback from the start, because retrofitting chargeback-compatible tagging after the fact is significantly more expensive than building it into the initial instrumentation design.

Diagram 12.3 — Cost Attribution: Chargeback vs Showback

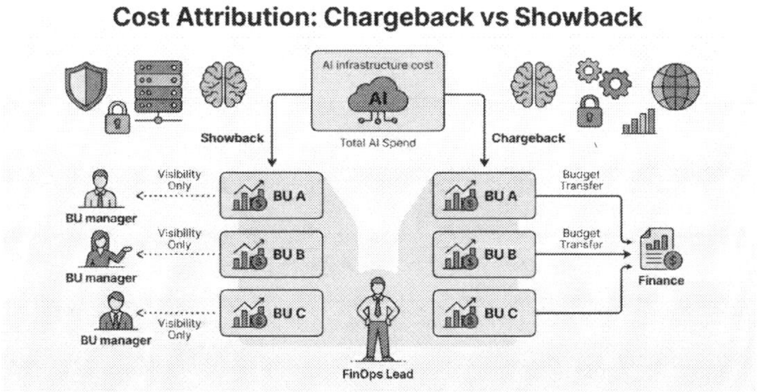

12.2 Optimization, Right-Sizing, and the FinOps Feedback Loop

Cost attribution tells the organization where its AI spending goes. Optimization converts that intelligence into action. The observability program is an engine of AI cost optimization because it provides usage, performance, and capacity utilization data that reveal where spending exceeds the value it generates. Right-sizing infrastructure, eliminating waste in serving configurations, and tuning model selection decisions are all optimization activities that require observability data as their primary input.

Right-Sizing AI Infrastructure

Right-sizing matches AI infrastructure capacity to actual workload requirements. Over-provisioning is endemic: model teams size their serving infrastructure for peak load at launch and never revisit that sizing as actual usage patterns emerge. A GPU inference cluster sized for 10,000 requests per second may be serving 800 requests per second six months later, with the remaining capacity idle and billed at the full hourly rate. At typical GPU instance costs, this pattern represents hundreds of thousands of dollars in annual waste per model.

The observability program provides the utilization data required for right-sizing: GPU utilization percentages, memory pressure, request queue depth, and latency distributions at different times of day. The ML platform engineer and FinOps lead review this data quarterly to identify over-provisioned models and evaluate whether the serving configuration can be right-sized by changing instance types, updating auto-scaling policies, or migrating to spot instances for latency-tolerant batch workloads.

Model-level right-sizing is separate from infrastructure right-sizing. For LLM workloads, the question is whether every use case requires the largest model in the approved catalog, or whether a smaller, cheaper model would produce acceptable quality at lower cost. Arize's evaluation capabilities and Langfuse's built-in scoring support systematic quality comparisons between model versions at different cost points. The FinOps lead should work with the ML platform lead to require evaluation of at least one smaller-model alternative before approving any new deployment to the most expensive model in the catalog.

Diagram 12.4 — Right-Sizing Decision Framework

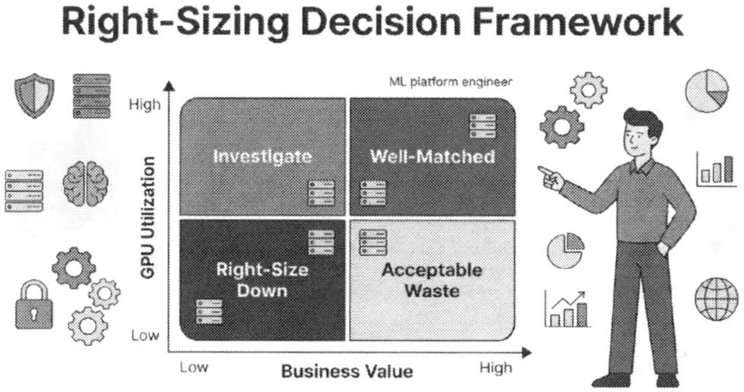

The Observability-Cost Feedback Loop

The most powerful cost optimization mechanism is the feedback loop between observability data and spending decisions. Without this loop, cost optimization is periodic: FinOps reviews the cloud bill quarterly and identifies anomalies afterward. With the loop, optimization is continuous: observability signals indicating rising cost-per-inference, declining utilization efficiency, or unexpected token consumption spikes trigger automated alerts that route to the FinOps lead and the relevant model team within hours of the anomaly.

Implementing the feedback loop requires that the observability platform captures cost-relevant metrics alongside operational metrics and applies the same alert infrastructure to cost anomalies. A sudden increase in average token count per LLM call may indicate that a prompt accidentally includes a large context block; an alert allows diagnosis before the issue incurs a material cost impact. A steady decline in GPU utilization on a serving cluster may indicate traffic has shifted to a different feature; an alert signals the infrastructure team to right-size before the cluster runs idle for another quarter.

The feedback loop also applies to the costs of observability infrastructure itself. An observability program that retains raw inference logs at full fidelity for all models incurs storage costs proportional to inference volume. The ML platform engineer should review observability infrastructure costs quarterly and apply the same right-sizing discipline to the monitoring stack that the FinOps lead applies to the serving stack. Sampling rate adjustments, tiered retention policies, and selective log aggregation are the primary levers for controlling observability costs without degrading signal quality.

Diagram 12.5 — Observability-Cost Optimization Loop

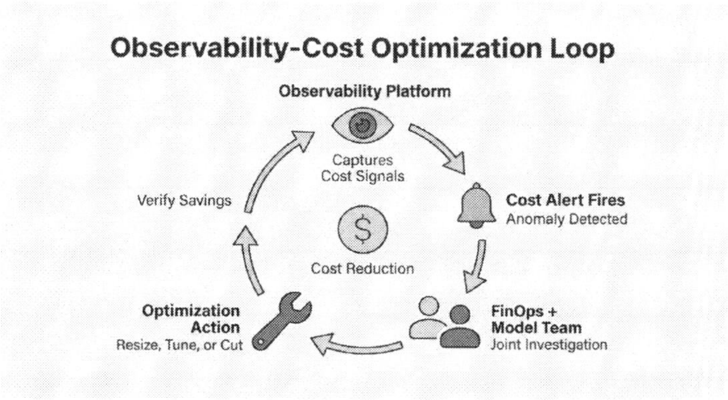

Budgeting and Forecasting AI Operational Costs

AI operational cost forecasting requires a model of the relationship between business activity drivers and infrastructure cost drivers. For LLM features, the primary driver is typically monthly active users and average queries per user; the cost model multiplies these by average tokens per query and the applicable token rate. For traditional ML serving, the driver is inference volume multiplied by cost per inference at the provisioned infrastructure level. Observability data provides the baseline measurements; the product team provides the growth assumptions.

The FinOps lead should build and maintain a cost model for each major AI cost center that supports scenario analysis: what happens to costs if user volume doubles, if the organization migrates to a newer model version with different pricing, or if average session length for a conversational feature increases by 50%? These scenarios are the inputs to the annual budget process and quarterly reforecast. Without them, the AI budget is a single-point estimate that cannot account for the inherent variance in AI usage patterns.

Cost forecasting also requires visibility into vendor pricing and the organization's contractual commitment positions. Most major cloud providers and LLM API vendors offer committed-use discounts of 20 to 40 percent in exchange for minimum spending commitments over one or three years. The FinOps lead should model the break-even point for each commitment option using current observability consumption data and present the analysis to the VP of AI and CFO as part of the annual budgeting cycle. An organization that can demonstrate consistent consumption at a specific level has a much stronger negotiating position with vendors than one that provides only monthly billing summaries.

Diagram 12.6 — AI Cost Forecast Model Components

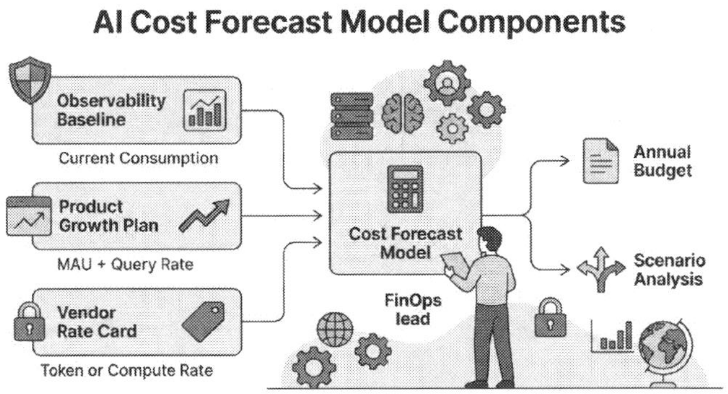

12.3 FinOps Frameworks and Vendor Negotiation for AI

The FinOps Foundation's framework for cloud financial management provides a useful structural model for AI cost management. Still, it requires adaptation because AI workloads have cost drivers that standard cloud FinOps does not address. The Inform, Optimize, and Operate phases of the FinOps cycle also apply to AI. Still, the specific practices within each phase must account for token economics, model version economics, and the dynamic pricing structures of LLM API vendors. This section describes the adaptations required and the vendor-negotiation strategies enabled by observability data.

Adapting the FinOps Framework for AI Workloads

The Inform phase requires that all stakeholders have visibility into AI costs at the granularity relevant to their decisions. For an ML platform engineer, the relevant granularity is per-model per-day utilization and cost. For a product manager, it is the cost per feature per active user. For a VP of AI, it is the total AI infrastructure cost as a percentage of AI-generated revenue. The observability program's cost attribution data must be surfaced in the FinOps dashboards at the granularity appropriate to each stakeholder's decision scope, not only at the billing account level visible in the cloud console.

The Optimize phase for AI requires practices beyond standard reserved instance optimization. AI-specific practices include model cascade routing, where requests are first evaluated by a cheap small model and escalated to a larger model only when the small model's confidence falls below a threshold; semantic caching, where results for semantically similar prompts are served without a new LLM call; and spot instance usage for latency-tolerant batch inference. These optimizations require coordination between the FinOps lead and the ML platform engineer, and their effectiveness should be measured through the observability platform's cost-per-inference metrics.

The Operate phase establishes the governance structures and review cadences that sustain cost discipline over time. For AI FinOps, the Operate phase includes: a monthly AI cost review meeting attended by the FinOps lead, the ML platform lead, and major consuming business unit representatives; a quarterly right-sizing review examining utilization data for all serving infrastructure; and an annual vendor commitment review assessing whether existing committed use discounts are optimally sized. These cadences mirror the governance-review structures from the previous chapter and should be combined where possible to reduce the meeting burden.

Diagram 12.7 — AI FinOps Cycle: Inform, Optimize, Operate

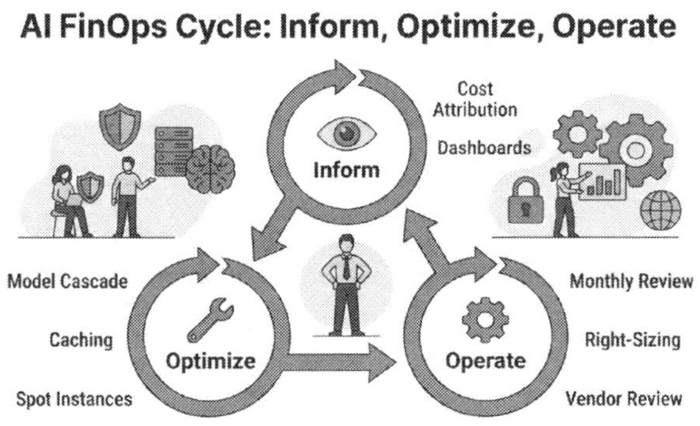

Negotiating with Cloud and AI Vendors Using Observability Data

Vendor negotiation for AI infrastructure is a management activity that most organizations approach reactively: the bill arrives, the VP asks for a discount, and the vendor offers a modest reduction in exchange for a commitment that may or may not align with actual usage. A FinOps lead who enters negotiations with detailed observability data is in a fundamentally different position. The data demonstrates consumption patterns, identifies peak and off-peak periods, and provides a credible forecast that the vendor can use to offer a committed use discount calibrated to the organization's actual trajectory.

For LLM API vendors, the primary negotiation lever is the committed use discount in exchange for a minimum monthly spend commitment. The FinOps lead should calculate the break-even monthly spend level using ninety days of observability-sourced token consumption data, add a conservative growth assumption, and use that number as the basis for the commitment offer. A commitment calibrated to the organization's actual consumption trajectory captures the full discount benefit without creating a commitment overhang, where the organization pays for committed capacity it does not consume.

For GPU compute vendors and cloud providers, negotiation levers include: on-demand versus reserved instance pricing for baseline capacity, spot instance availability commitments for batch workloads, and custom pricing tiers for organizations above a specific annual spend threshold. The FinOps lead should prepare a consumption report from the observability platform before each major negotiation, showing the vendor three to six months of actual consumption by service category, the growth trend, and the projected consumption for the next contract period. Vendors respond to evidence of demonstrated consumption because it reduces their revenue-forecasting uncertainty.

Diagram 12.8 — Vendor Negotiation Evidence Package

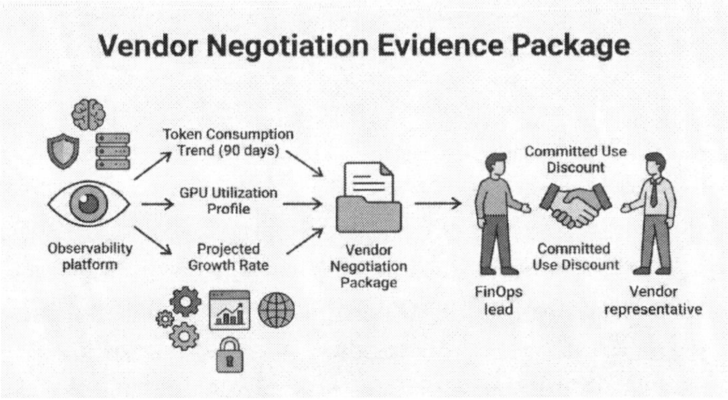

Making Observability Investment Self-Funding

A recurring objection to observability investment is its cost: at the scale of a large enterprise AI portfolio, observability infrastructure costs can reach seven to twelve percent of total AI infrastructure spending. The response requires demonstrating that observability investment generates returns exceeding its cost through cost optimization, incident prevention, and compliance preparation benefits.

The cost optimization case is the most direct. An observability program that enables a single right-sizing decision for an over-provisioned GPU cluster typically saves more in one quarter than the program's annual infrastructure cost. An LLM token optimization project reducing the average token count by 20% for a high-volume feature generates continuous savings that compound with usage growth. A vendor commitment negotiation that captures a 30% discount on a $2 million annual API spend saves $600,000 per year. Each outcome requires the observability data that the program produces and exceeds the program's operating cost by a large margin.

The compliance and incident prevention cases are less immediately quantifiable but often larger in dollar terms. The cost of a single SR 11-7 matter-requiring-attention citation, including remediation work and halted model deployments, typically exceeds two to three years of operating costs for the observability program. A single major AI incident triggering a regulatory investigation can cost orders of magnitude more. The observability lead should present this self-funding argument annually to the VP of AI and the CFO, using the previous year's cost savings, incident-prevention outcomes, and governance outcomes as evidence.

Diagram 12.9 — Observability ROI: Cost vs Benefit

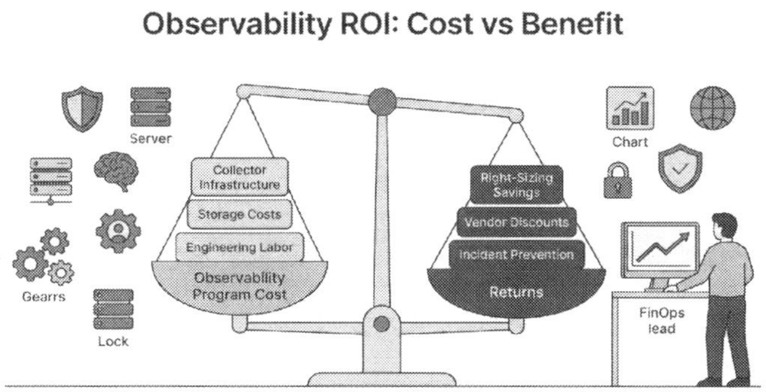

12.4 Manager's Checklist

- Build a total AI cost of ownership inventory that includes compute, storage, token API fees, data transfer, and engineering labor, not only the cloud compute bill.
- Instrument every LLM inference call to capture prompt and completion token counts tagged by feature identifier and business unit before any new LLM feature reaches production.

- Implement cost attribution at the model level as a minimum and progress toward feature-level attribution within twelve months of first production LLM deployment.
- Review GPU utilization data quarterly for all serving infrastructure and right-size any cluster running below fifty percent average utilization.
- Establish a monthly AI cost review meeting attended by the FinOps lead, ML platform lead, and major consuming business unit representatives.
- Prepare a consumption report from the observability platform before every major vendor commitment negotiation, covering ninety days of actual usage by service category.
- Build an LLM cost model for each high-volume feature that expresses cost in unit economics terms, such as cost per resolved ticket or cost per user query answered.
- Present an annual observability self-funding analysis to the VP of AI and CFO, showing cost savings, incident prevention value, and compliance preparation benefits against program operating costs.

12.5 Takeaway

AI infrastructure costs are not self-managing. Without deliberate observability instrumentation that captures token counts, utilization rates, and feature-level consumption alongside operational performance signals, AI spending becomes opaque, and the organization loses the ability to link infrastructure costs to business value at the granularity required for sound investment decisions. The FinOps lead and the ML platform lead must operate as partners, with the observability program as the shared data foundation for every cost-attribution, optimization, and vendor-negotiation decision they make.

The observability program pays for itself many times over through the cost savings it enables. Right-sizing decisions, token optimization projects, and vendor commitment negotiations each generate returns that compound as the AI portfolio grows. The discipline of treating AI infrastructure cost with the same rigor as AI operational quality is not a finance team responsibility grafted onto a technical program; it is a core management capability that determines whether the organization's AI investment generates a durable competitive advantage or accumulates as an uncontrolled overhead burden.

13 Building the AI Observability Team

A regional insurance carrier had deployed seven predictive models into production over 18 months. The ML team that built them had long since moved on to new projects. Incidents were caught by customer complaints, not by monitoring dashboards. When the state insurance regulator requested a model performance audit, the VP of AI/ML discovered that no single person owned the observability function. The data science team said it was an engineering concern. Platform engineering said it was a data science concern. Operations said it was both. The carrier spent eleven weeks and roughly $400,000 reconstructing audit artifacts that a properly staffed observability team would have maintained as routine. That experience is not unusual. It is the predictable result of building an AI production capability without also building the human infrastructure to watch over it.

AI observability does not run itself. Every dashboard, every alert threshold, every drift detection job, and every post-incident review depends on people who own those responsibilities and carry the skills to execute them. Most enterprises staff their AI programs heavily on the build side and then treat production oversight as an afterthought absorbed by whoever has the most available capacity. That gap produces exactly the kind of invisible degradation described in earlier chapters. Closing it requires deliberate decisions about roles, reporting lines, skill requirements, and career paths.

The organizational stakes are direct. A mis-staffed observability function creates regulatory exposure when audit requests arrive, and no one can produce a coherent performance history. It creates budget risk when token costs escalate without a FinOps-fluent engineer watching the meters. It creates operational risk when a drifted fraud-detection model continues to generate decisions for weeks because no one is accountable for the weekly drift review. Talent decisions in this domain are governance decisions. A VP of AI/ML who treats observability headcount as optional overhead is flying blind on every model in production.

13.1 The Skills Gap and Why It Persists

Most organizations hiring for AI roles optimize for model-building talent: data scientists and ML engineers focused on architecture and feature engineering. That talent is valuable, but it is not the talent that keeps production AI systems healthy. The skills required to instrument a model pipeline, design a drift-detection regime, or write a runbook for an LLM guardrail failure are a distinct combination of MLOps, software reliability engineering, and data-quality disciplines. Because no university program yet produces graduates with this precise profile, most enterprises either discover the gap after something breaks or never recognize it at all.

What AI Observability Actually Requires

The skill profile for effective AI observability sits at the intersection of three disciplines. The first is traditional software reliability engineering: distributed tracing, log aggregation, metric cardinality, and alert routing. The second is ML engineering: reading model performance curves, interpreting confusion matrices, understanding feature importance, and knowing when a statistical signal indicates meaningful degradation rather than natural variance. The third is data quality engineering: validating schema contracts, monitoring ingestion freshness, detecting upstream pipeline corruption, and maintaining lineage records from source to prediction.

Few professionals arrive with deep expertise in all three. A platform engineer who has mastered Datadog and OpenTelemetry will need structured exposure to model-quality concepts. A data scientist comfortable with drift statistics may have never written an on-call runbook or configured a PagerDuty escalation policy. A data engineer who understands Great Expectations validation may be unfamiliar with the semantic evaluation frameworks that LLM observability tools like Langfuse and Arize use to score output quality. Recognizing this three-discipline gap is the first management act. Pretending that one discipline covers the others yields teams that capture only a fraction of the failure modes of their production systems.

The gap persists for structural reasons. Data science teams are typically measured on model accuracy improvements and feature delivery velocity, not on production health. Observability work is invisible when it succeeds and blamed on other functions when it fails. Without explicit role definitions, performance metrics tied to production health, and career paths that treat observability as a senior specialty, talented engineers tend to prioritize work that earns recognition over observability. Managers who want to close this gap must address those structural incentives directly.

Diagram 13.1 — Three-Discipline Skills Map

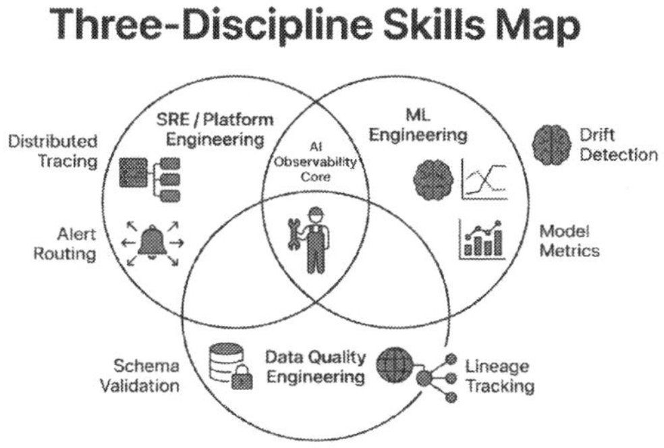

The Hiring Market Reality

Job boards today show thousands of postings for ML engineers and data scientists, and a fraction of that volume for roles titled AI SRE, MLOps Engineer, or Model Observability Engineer. That disparity reflects the market's emphasis on AI creation over AI operation. Candidates who can credibly claim expertise across distributed tracing, model performance monitoring, and LLM evaluation frameworks command salaries competitive with senior data science roles, and there are far fewer of them available.

Organizations that have successfully built observability teams tend to grow them internally rather than hire them wholesale. They identify SREs who are curious about ML systems and invest in structured upskilling. They identify data scientists who are drawn to production engineering problems and give them rotational time in platform roles. They hire one or two senior external hires to set technical direction and mentor internally grown talent. This hybrid approach takes twelve to eighteen months to produce a functioning team, which means the hiring decision must be made before the observability gap becomes a crisis.

A manager planning team-building in this space should treat the external hiring plan and the internal development plan as equally important work streams. Posting a job requisition and waiting for the perfect candidate is not a strategy. Identifying the two or three engineers on the current team most likely to grow into observability roles and designing a structured development plan for each is the more reliable path to coverage within a fiscally reasonable timeframe.

Diagram 13.2 — Build vs. Hire Staffing Funnel

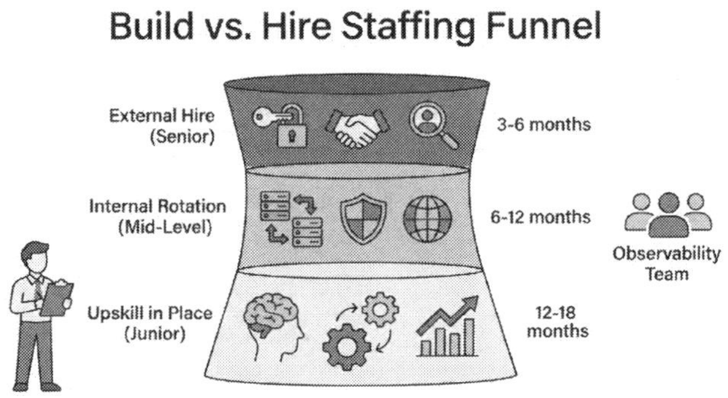

Framing Observability as a Career Specialty

Retention of observability talent depends on whether the organization treats the function as a genuine specialty with its own career ladder or as a support role subordinate to data science. An engineer with expertise in OpenTelemetry instrumentation, PSI-based drift detection, and LLM output evaluation using Arize or Fiddler can find a home at any enterprise with a serious AI program. If your organization offers no title progression beyond a generic MLOps Engineer role, that engineer will eventually leave for a place that does.

Building a specialty track means defining at least three levels: a practitioner role focused on instrumentation, alerting, and runbook execution; a senior role focused on architecture, threshold design, and cross-team standards;

and a staff or principal role focused on platform strategy, vendor evaluation, and observability program governance. Each level should have explicit skill criteria that distinguish it from the level below. The senior role might require demonstrated ability to design a statistical sampling strategy for LLM evaluation. In contrast, the staff role might require experience in presenting the health of the observability program to an audit committee.

When observability carries a real career path, it attracts engineers who want to build a long-term specialty rather than treat the role as a stepping stone. Those engineers invest in the function, stay current on emerging tools like the OpenTelemetry GenAI semantic conventions, contribute to vendor evaluations, and build the institutional knowledge that makes a team resilient when individuals leave. Career ladder design is not HR formality. In this domain, it is a retention-and-program-quality strategy.

Diagram 13.3 — Observability Career Ladder

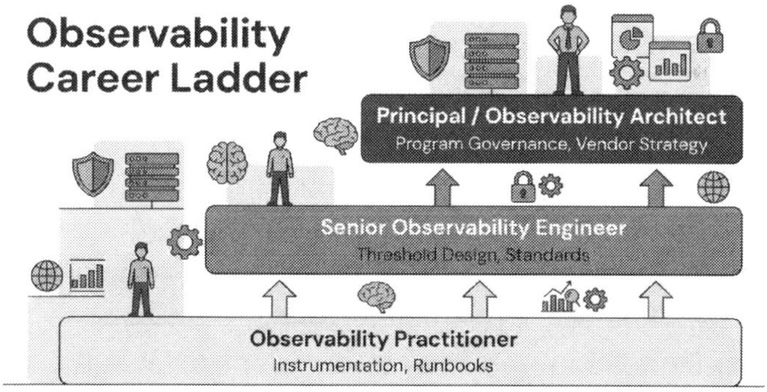

13.2 Role Definitions and Organizational Models

Clear role definitions are the management artifact that converts a vague commitment to observability into a functional program. Without them, tasks fall through gaps between teams, accountability is diffuse, and when an incident occurs, the first conversation is about who should have been watching rather than what to do. This section maps the five core roles of a mature AI observability function and describes three organizational models enterprises have used to organize those roles.

The Five Core Roles

The ML Platform Engineer owns the infrastructure on which models run: training pipelines, inference endpoints, model registries, and the compute layer that executes predictions. In an observability context, this role is responsible for instrumenting the platform with OpenTelemetry collectors, configuring GPU utilization and latency metrics, and ensuring that trace data flows from inference endpoints into the observability backend. The Data Quality Engineer owns the data that flows into models. This role defines schema contracts using tools like Great Expectations or Deequ, monitors ingestion freshness and completeness, and tracks feature distributions against training baselines.

The AI SRE bridges the infrastructure and model quality layers, focusing on service-level objectives for model-dependent applications: response time, prediction throughput, error rates, and the business impact of model failures. The AI SRE writes incident runbooks, owns the on-call rotation for model-related alerts, manages escalation paths in PagerDuty or ServiceNow, and leads post-incident reviews. The Observability Lead sets strategy, owns the observability platform, evaluates and manages vendor relationships with tools such as Langfuse, Arize, and Datadog, and translates technical signals into management reporting.

The AI Governance Lead connects observability outputs to compliance requirements, ensuring that audit artifacts meet regulatory standards and that model performance records are maintained with the retention and access controls required by frameworks such as the EU AI Act. This role is most frequently absent from early team builds, and that absence becomes the most costly gap when regulatory inquiries arise. It requires a different orientation from the operational roles: comfort with policy language and audit processes, alongside enough technical fluency to understand what the observability data actually represents.

Diagram 13.4 — Five Core Observability Roles

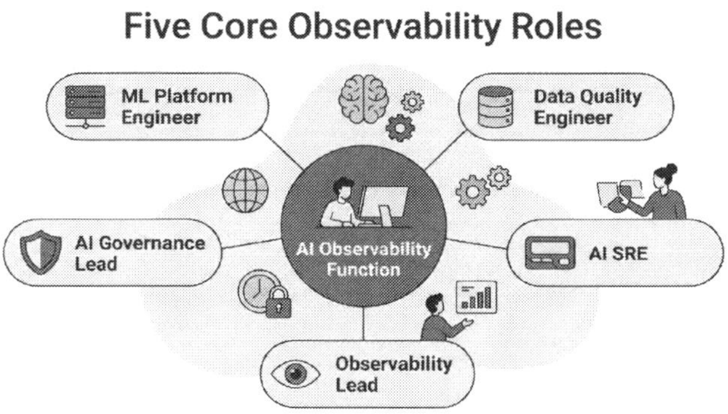

Organizational Models: Embedded, COE, and Hybrid

The embedded model places observability engineers directly within product or domain teams. The advantage is deep domain knowledge and tight coupling with model owners. The disadvantage is duplication of effort, inconsistent standards across teams, and the near-impossibility of building specialized expertise when each engineer is the only observability resource in their domain. Embedded models work reasonably well when an organization has fewer than five production AI applications, each with distinct infrastructure and no shared platform.

The centralized center-of-excellence model creates a dedicated observability team that serves all AI product teams from a shared platform. This team owns the observability tooling, sets instrumentation standards, operates the alerting infrastructure, and provides consulting services to product teams. The advantage is consistency, specialization, and the ability to spot cross-portfolio patterns that embedded teams would never see. COE models work best in organizations with 10 or more production AI systems sharing a common infrastructure, where the breadth of coverage justifies the investment in a centralized team it provides.

The hybrid model combines a small central team, typically an Observability Lead and one or two senior engineers who own the platform and set standards, with observability-accountable roles embedded in product teams. Those embedded roles are ML engineers or data engineers who carry an explicit observability accountability as part of their job description and participate in a cross-functional observability community of practice. This model is the most common choice for mid-sized enterprises with five to fifteen production AI applications, capturing the standards benefits of the COE model without the headcount of a fully centralized function.

Diagram 13.5 — Three Organizational Models Compared

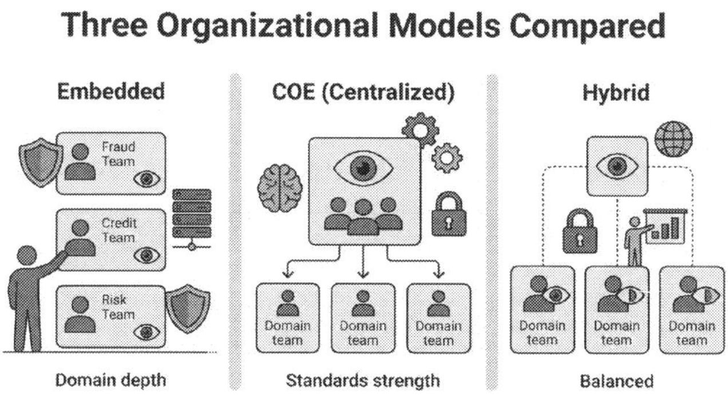

Governance Lead and Executive Sponsorship

The AI Governance Lead should not be the same person as the Observability Lead or the AI SRE. Those roles focus on operational health; the governance role focuses on evidence and accountability. In smaller organizations, this role may be allocated at 50 percent to a senior data scientist or a risk officer who receives structured training in AI observability concepts, rather than to a dedicated full-time hire. The connection between the observability program and the enterprise's GRC platform, whether ServiceNow GRC, Archer, or a custom solution, runs through this role.

Executive sponsorship is distinct from the governance role but equally critical. Without a VP of AI/ML or a CIO who visibly advocates for the observability program, allocates budget to it, and references it in business reviews, more visible priorities will always squeeze out the function. The executive sponsor does not need to understand every technical detail. They need to understand three things: what the observability program is watching, what it costs, and what the enterprise would be exposed to without it.

A one-page program summary reviewed quarterly by the sponsor, covering portfolio health, open incidents, upcoming regulatory touchpoints, and platform costs, is a simple artifact that keeps sponsorship active and informed. Organizations that formalize this reporting cadence find that budget defense at planning time becomes significantly easier. The sponsor has current data, speaks from knowledge, and can directly counter the argument that observability is overhead rather than an essential operating cost.

Diagram 13.6 — Executive Sponsorship Accountability Chain

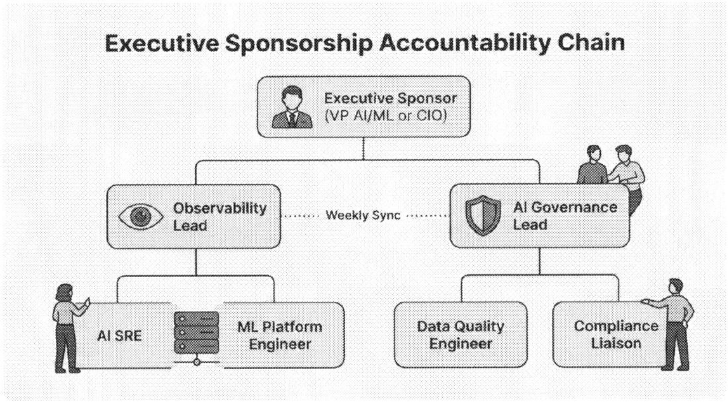

13.3 Cross-Functional Collaboration and Upskilling

An observability team with clear roles and the right org model still fails if it cannot collaborate effectively across the functions that produce and consume AI systems. Data science teams resist additional instrumentation requirements. Security teams want access to observability data that the ML team considers proprietary. Compliance teams arrive with audit requests that the observability team has not anticipated. Product owners want dashboards that tell a business story, not a statistical one. Managing these interfaces requires structured agreements, not goodwill.

Bridging Data Science, Engineering, and Compliance

The most common friction point between observability teams and data science teams is instrumentation ownership. Data scientists view the model as their artifact; they built it, validated it, and are accountable for its performance. When an observability engineer proposes adding logging hooks to a model-serving endpoint, data scientists sometimes perceive it as an implicit critique of their work rather than a production necessity. The manager's role is to establish, clearly and early, that instrumentation is a prerequisite for deployment. Models that cannot be observed do not go to production. That standard, enforced uniformly, removes the friction from the individual relationship and places it at the policy level where it belongs.

Engineering teams that own the CI/CD pipeline need to understand that AI workloads require additional telemetry beyond what traditional application monitoring provides. An OpenTelemetry collector configured for a web API does not automatically capture feature distribution data or model confidence scores. A joint working session between the ML Platform Engineer and the DevOps team, focused specifically on what additional spans and attributes the AI observability instrumentation requires, is a necessary early investment. The output should be a shared instrumentation standard that both teams own and maintain.

Compliance teams are often late entrants to the observability conversation, arriving only when an audit is imminent. The better pattern is to include the compliance team in the initial program design by asking what evidence they would need to demonstrate that a high-risk model is being monitored appropriately. That question surfaces record retention requirements, access control requirements, and reporting format preferences that the technical team would otherwise discover at the worst possible moment. Building compliance requirements into the observability architecture from the start is far cheaper than retrofitting them under audit pressure.

Diagram 13.7 — Cross-Functional Collaboration Map

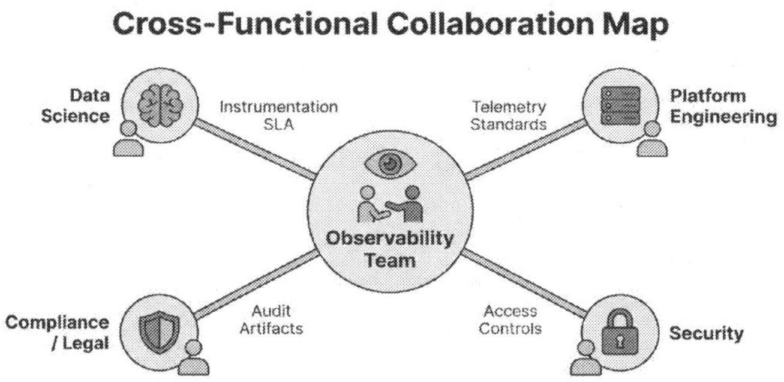

Upskilling Paths from Traditional Ops

The fastest route to a staffed observability team is converting existing SRE and operations talent rather than hiring from scratch. Traditional SREs already understand distributed tracing, log pipelines, metric cardinality, and on-call rotation management. The gap is in ML-specific knowledge: understanding what PSI measures, how to interpret a precision-recall tradeoff, what a confidence calibration curve reveals about a model's reliability, and how LLM evaluation frameworks like RAGAS or DeepEval score faithfulness and relevance. A structured twelve-week curriculum can close the core of that gap for an experienced SRE.

The curriculum should be organized into three phases. The first four weeks focus on ML fundamentals: supervised and unsupervised learning, common model types, feature engineering basics, and the statistical foundations of drift detection. Practical exercises should be grounded in the organization's actual models and use real training data distributions. The second four weeks focus on observability tooling: hands-on configuration of Arize, Evidently, or Langfuse in a staging environment; writing custom metrics; building dashboards; and setting alert thresholds with documented justification. The third four weeks focus on governance and incident response: reading and writing model cards, conducting a mock regulatory inquiry, and completing a full AI incident response simulation from detection through post-incident review.

Organizations that have run this curriculum report that experienced SREs typically outperform external hires by month six, because they already have deep knowledge of the production environment, existing relationships with engineering teams, and credibility with the on-call rotation. The investment is twelve weeks of structured learning time and access to a senior ML engineer or Observability Lead who serves as a mentor. Treating this investment as overhead is a mistake; it is the fastest, most cost-effective way to build a credible observability function.

Diagram 13.8 — SRE-to-AI Observability Training Path

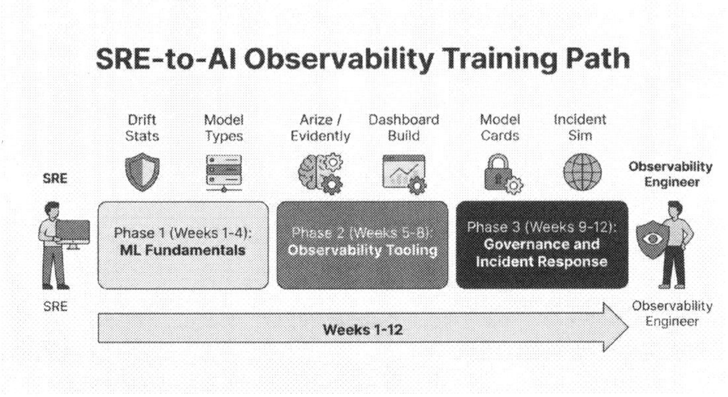

Vendor Management and the Observability Champion

Every production observability program involves vendor relationships that require active management. An observability platform contract with Arize, Datadog, or Dynatrace may represent $200,000 to $2,000,000 in annual spend at enterprise scale. Without a designated owner for that relationship, the contract renews on default terms, feature adoption stagnates at whatever was configured at go-live, and the vendor's support team becomes the de facto system administrator for your observability infrastructure.

The Observability Lead should own the primary vendor relationship and conduct a structured quarterly review with the vendor's customer success team. That review should cover three topics: utilization against contracted capacity; new platform capabilities the organization has not yet adopted; and any reliability or support incidents from the prior quarter. The quarterly review is also the right venue for discussing renewal terms before contract expiry creates time pressure.

The executive champion, whether a VP of AI/ML, a Chief Data Officer, or a CIO, should receive a brief monthly report covering three numbers: the count of active production models under monitoring coverage, the number of incidents detected and resolved in the prior month, and the current observability platform cost per monitored model. Those three numbers tell an executive everything needed to defend the program budget in a planning cycle. A program without an executive champion is always one budget cycle away from losing its headcount.

Diagram 13.9 — Vendor Review and Champion Cadence

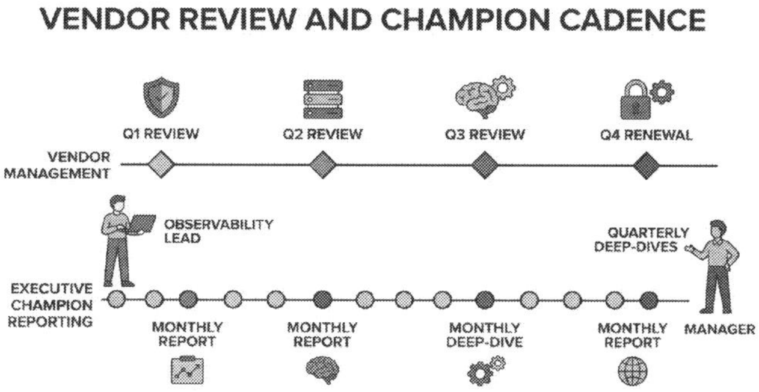

13.4 Manager's Checklist

- Define all five observability roles in writing before posting any requisitions, including explicit skill criteria for each level.
- Choose an organizational model (embedded, COE, or hybrid) and document the rationale before the first hire.
- Require instrumentation as a deployment gate before any model is promoted to production; enforce this uniformly across all teams.
- Identify two or three internal SRE or data engineering candidates for upskilling and design a twelve-week development plan for each.

- Schedule a joint instrumentation standards session between the ML Platform Engineer and the DevOps team within the first thirty days.
- Invite the compliance team to the observability program design review and capture their audit evidence requirements before the architecture is finalized.
- Conduct a quarterly vendor review with every observability platform provider covering utilization, new capabilities, and renewal terms.
- Establish a monthly one-page observability program summary reviewed by the executive champion, covering portfolio coverage, incidents, and cost per model.

13.5 Takeaway

Building an AI observability team is not a staffing exercise. It is a governance decision that determines whether your organization can see what its AI systems are doing in production. The skills gap is real, the roles are distinct, and the organizational model requires deliberate choice rather than default drift. Every week that passes without clear accountability for production AI health is a week in which silent degradation accumulates undetected.

A hybrid model with a small central team and observability-accountable embedded engineers can cover a portfolio of ten to fifteen production models with six to eight people who have the right training, the right tools, and the right management support. The return is operational clarity: knowing exactly how each production AI system is performing, what the trends are, and who will respond when something goes wrong.

14 The AI Observability Maturity Model

A global logistics company had been running AI-powered route optimization for three years. Their ML team celebrated every model release, meticulously tracked training accuracy, and maintained a model registry in Databricks with clean version control. What they lacked was any systematic view of how those models performed once deployed. Alerts are fired only when the serving endpoint returns a 500 error. Drift was discovered quarterly when analysts noticed on-time delivery rates had slipped. When the company benchmarked against a peer that had invested in a proper observability program, the gap was stark: the peer detected and resolved model degradation events four times faster at a fifth of the incident volume. The difference was not technology. It was maturity. The logistics company was at Level 2; the peer was at Level 4.

A maturity model gives managers a shared language for assessing where their observability program actually stands, not where they believe it stands based on the technology they have purchased. The gap between those two assessments is almost always significant. Organizations routinely overestimate their maturity by conflating having a tool with having a practice. A Datadog license does not make an organization Level 3. Actual maturity is a function of people, process, and technology working together in a disciplined way.

The practical value of the maturity model is that it converts an abstract aspiration, improving our AI monitoring, into a concrete roadmap with specific milestones. A manager who can say precisely which Level 3 capabilities are missing, what it would take to acquire them, and what the business risk of remaining at Level 2 costs in terms of delayed incident detection or compliance exposure is positioned to make a credible budget request and a credible commitment to leadership. The maturity model is both a diagnostic and a planning instrument.

14.1 The Five Maturity Levels Defined

The five-level AI observability maturity model describes progression from complete blindness to predictive, AI-driven oversight. Each level is defined by what the organization can detect, how quickly it can respond, and what decisions it can confidently make based on observability data. The levels are assessed against specific, concrete criteria. An organization either has the capability, or it does not, and the assessment should be made by reviewing actual artifacts, not by polling team members.

Level 1 Reactive and Level 2 Basic

At Level 1, the organization has no intentional AI monitoring program. Model failures are discovered by end users who observe incorrect outputs, by downstream systems that produce anomalous results, or by business analysts who notice KPI deterioration during monthly reviews. There are no alert thresholds on model endpoints, no feature distribution tracking, and no formal process for distinguishing a model quality problem from an infrastructure problem. The mean time to detect a model degradation event at Level 1 is measured in weeks or months, because detection depends entirely on the lag between model output and observable business outcome.

Level 2 represents the first deliberate step: infrastructure metrics and uptime monitoring are in place, but they cover system health rather than model quality. The team can tell whether the inference endpoint is responding, whether GPU utilization is within normal bounds, and whether API error rates are elevated. Tools such as Datadog, New Relic, and Prometheus collect these metrics. What Level 2 cannot detect is the subtle, more consequential class of failures: a fraud model whose precision has declined from 91 percent to 78 percent over six weeks; a credit scoring model whose feature distributions have shifted following a data pipeline change; an LLM application whose response quality has deteriorated because a prompt template was updated without evaluation.

The critical management insight at Levels 1 and 2 is that the investment to move from Level 1 to Level 2 is primarily a technology investment, whereas moving from Level 2 to Level 3 is primarily an organizational investment: defining what model quality means for each production system, establishing baselines, and assigning accountability for reviewing them. Organizations that make the technology investment without organizational investment tend to stall at Level 2 for extended periods, maintaining the appearance of a monitoring program without the substance.

Diagram 14.1 — Level 1 vs Level 2 Detection Gap

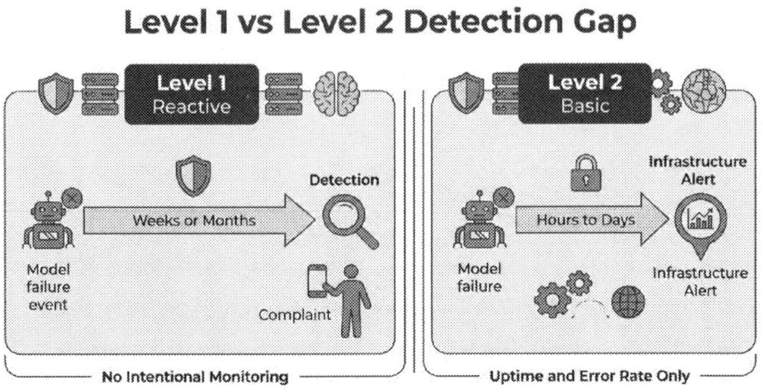

Level 3 Proactive and Level 4 Advanced

Level 3 introduces model performance tracking: the organization monitors accuracy, precision, recall, or business-aligned proxy metrics for each production model and runs drift detection on a scheduled basis. Feature distribution comparisons using the Population Stability Index or the Kolmogorov-Smirnov tests at daily or weekly cadence. The team has defined meaningful drift thresholds for each model, and alerts fire when those thresholds are crossed. Tools like Evidently, Arize, or Fiddler are deployed with model-specific monitoring profiles. Level 3 is the minimum viable observability state for any organization running high-stakes AI, such as models influencing credit decisions, healthcare triage, or fraud adjudication.

Level 4 represents the transition from reactive alerting to integrated, automated governance. At Level 4, evaluation is continuous: real-time or near-real-time drift detection runs on streaming feature data, LLM outputs are evaluated automatically using judge-model frameworks available in Arize or Langfuse, and alert policies are integrated with ITSM platforms like ServiceNow or PagerDuty. The observability program is connected to the model registry: when a new model version is promoted, baseline performance profiles are automatically imported into the monitoring system. Governance artifacts, including model cards and validation reports, are generated automatically and stored in a format accessible to audit teams.

The organizational characteristic that defines Level 4 is cross-functional integration. The observability program is not a separate function that produces reports for other teams; it is embedded in the deployment workflow, the compliance review process, and the executive reporting cadence. The AI SRE is on the same escalation path as the platform SRE. The AI Governance Lead reviews observability outputs in the quarterly compliance review. The VP of AI/ML sees a portfolio health dashboard alongside the product roadmap in every business review.

Diagram 14.2 — Level 3 to Level 4 Capability Jump

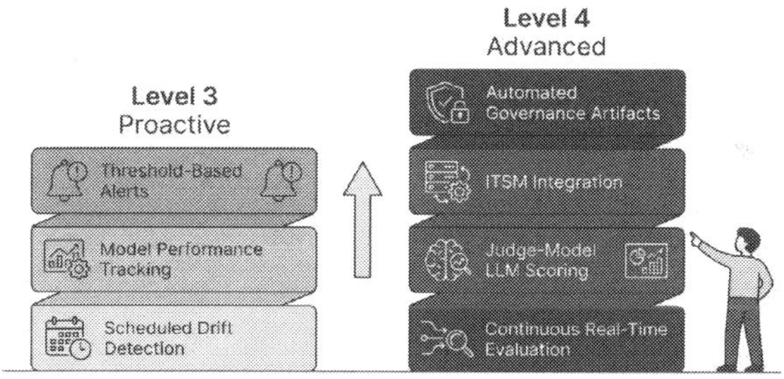

Level 5 Predictive

Level 5 represents the frontier: the organization uses AI to observe AI. Anomaly detection models continuously analyze the telemetry streams produced by production models, identifying degradation patterns before they exceed alert thresholds. Remediation workflows are partially automated: when a classification model's PSI score exceeds a warning threshold, a retraining pipeline is triggered automatically, and the resulting candidate model is queued for human validation rather than requiring an engineer to initiate the process manually. Cost optimization signals feed directly from the observability platform into the FinOps workflow, flagging underutilized inference endpoints for right-sizing.

At Level 5, the observability platform produces forward-looking signals rather than only backward-looking reports. Trend analysis projects when a model's performance is likely to cross a critical threshold given its current degradation rate, giving the team lead time to intervene before a business impact occurs. LLM applications are monitored at the semantic level: not just whether keyword filters flag responses, but also whether the distribution of topics, intent categories, and sentiment in user interactions is shifting in ways that suggest a fundamental change in how users engage with the system.

Level 5 is not a destination most enterprises will reach within a two-year planning horizon, and that is acceptable. The value of defining Level 5 is not to create an obligation but to give advanced organizations a clear direction for investment once Level 4 capabilities are fully operational. Organizations that spend resources chasing Level 5 capabilities before they have disciplined Level 3 practices are making a prioritization error. The maturity model enforces sequencing: each level depends on the foundations laid by the previous one.

Diagram 14.3 — Level 5 Predictive Observability Loop

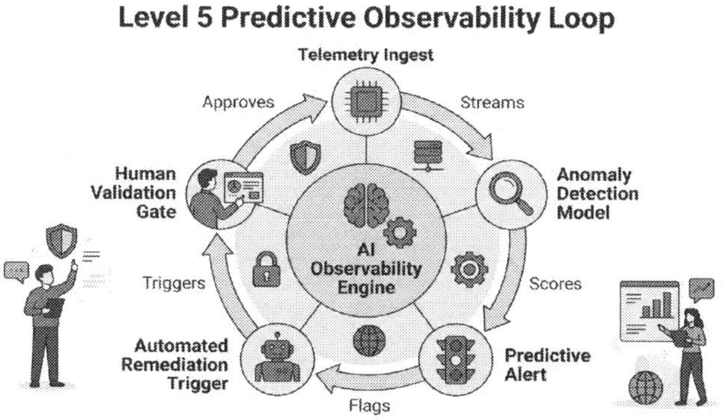

14.2 Self-Assessment and Gap Analysis

Knowing that five maturity levels exist is not the same as knowing which level your organization occupies. Self-assessment converts the model from a taxonomy into an action plan. It requires honest evaluation against artifact-based criteria: not what the team believes it can do, but what it can demonstrate it does regularly. A well-run self-assessment produces a gap analysis that identifies the specific capabilities missing at the next maturity level.

The Self-Assessment Checklist by Level

Each maturity level carries specific criteria that an organization must satisfy to claim it. Level 2 requires that infrastructure metrics, specifically CPU and GPU utilization, memory, endpoint latency, and API error rates, be collected, retained for at least thirty days, and tied to alerts with defined escalation paths. If those conditions are not fully met, the organization is at Level 1 regardless of what technology it has purchased. Level 3 requires, in addition to all Level 2 criteria, that model-specific performance metrics be tracked for every production model, that feature distribution baselines be established and compared on a documented schedule, and that a defined process exists for investigating and resolving drift alerts.

Level 4 criteria add: continuous or real-time evaluation for at least high-risk production models; integration of model quality alerts with the enterprise ITSM platform; automatic baseline import when new model versions are promoted; and governance artifacts generated as part of the deployment workflow rather than produced manually upon request. Level 5 criteria add: anomaly detection running over observability telemetry streams; automated remediation triggers for at least one well-understood degradation class; and forward-looking trend analysis producing predicted time-to-threshold estimates.

The checklist should be evaluated against evidence, not assertions. For each criterion, the assessor should ask: What artifact demonstrates that this capability is operational? For Level 3 drift detection, the answer might be the alert history showing that drift alerts fired and were investigated in the past ninety days. For Level 4 governance artifacts, the answer might be deployment pipeline logs showing that model cards were automatically generated during the last three model promotions. Where no artifact exists, the capability does not exist, and the gap is recorded for the roadmap.

Diagram 14.4 — Maturity Self-Assessment Scorecard

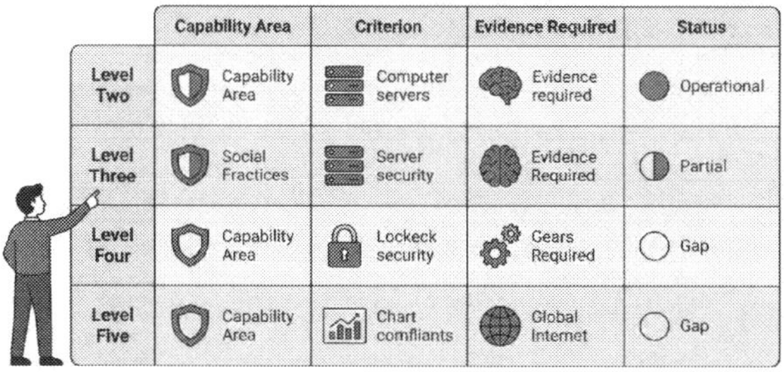

Common Gap Patterns by Organization Type

Different types of organizations exhibit predictable patterns of gaps. Financial services organizations typically have strong Level 2 infrastructure monitoring, inherited from mature DevOps practices, but stall at Level 3 because the quantitative analyst communities that own model validation have not been integrated into production monitoring workflows. The gap is organizational, not technical: the tools for Level 3 monitoring have been purchased. Still, the people who understand the models do not own the monitoring configuration, and those who own the monitoring do not understand the models well enough to set meaningful drift thresholds.

Healthcare organizations frequently have the opposite pattern: strong awareness of model quality concerns driven by FDA guidance on AI-based software, but weak infrastructure monitoring because healthcare IT has historically prioritized application uptime over observability depth. They reach Level 3 in model quality tracking, but gaps in the infrastructure layer limit their ability to correlate model degradation with upstream data pipeline failures. Retail and e-commerce organizations tend to reach Level 3 and partial Level 4 capabilities faster due to their existing real-time data infrastructure, but often have weak governance artifact production because compliance pressure has historically been lower in that sector.

Manufacturing and industrial organizations running AI on operational data face a distinct challenge: many of their models operate on sensor data from proprietary systems not designed for modern observability tooling. Their gap is often in the data collection layer, which limits feature distribution tracking even at Level 3. Recognizing these sector-specific patterns helps managers calibrate their roadmap against peer benchmarks rather than building a plan based solely on internal assessments, which tend to be overly optimistic.

Diagram 14.5 — Sector Gap Pattern Heatmap

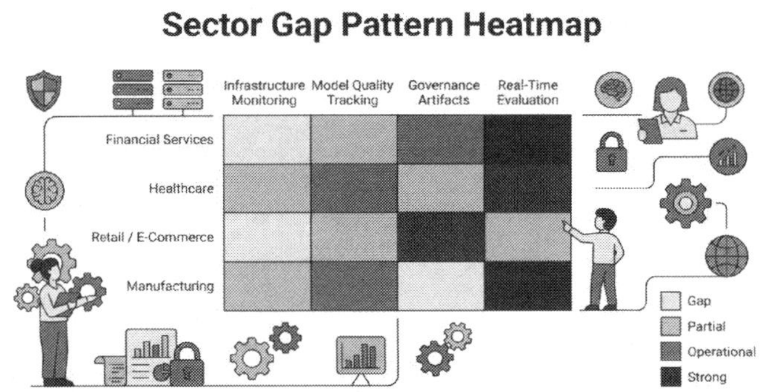

Translating the Gap Analysis into a Business Case

A gap analysis that stays inside the observability team is a missed opportunity. The real value of the self-assessment is the business case it enables for investment. That business case has three components. The first is the cost of the current gap: what incidents has the organization experienced in the past twelve months that better observability would have detected earlier, and what did those incidents cost in remediation time, business impact, and regulatory exposure? Even a rough estimate grounded in actual incident records is more persuasive than a theoretical argument.

The second component is the investment required to close the priority gaps: what technology, headcount, and process changes are needed to move from the current level to the next, and what does that cost over twelve months? This estimate should distinguish capital components, typically tool licensing, from operational components, typically engineering time allocated to configuration and ongoing management. The third component is the risk-adjusted return: given the incident frequency and business impact observed in the past year, what is the expected reduction in incident cost at the next maturity level, and how does that compare to the investment required?

A business case framed this way is legible to a CFO and to a board audit committee, not just to a technology leadership team. It connects observability investment directly to business risk reduction and expresses that connection in financial terms. Managers who bring this framing to budget discussions report significantly higher approval rates for observability program investments than those who frame the request in purely technical terms. Mission-aligned observability investment is easier to fund when the mission is expressed in the language of risk and return, not just in the language of system reliability.

Diagram 14.6 — Observability Investment Business Case

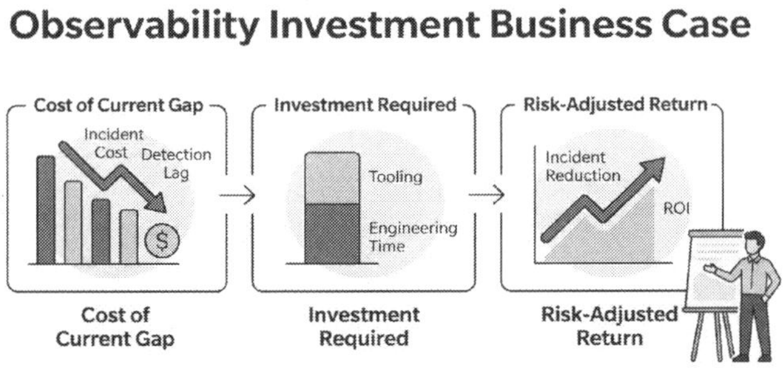

14.3 Roadmaps, Quick Wins, and Implementation Discipline

A maturity model without an implementation plan is an academic exercise. This section translates the self-assessment output into three planning horizons: a thirty-day sprint targeting immediate quick wins; a ninety-day roadmap targeting the capabilities needed to reach the next maturity level; and a twelve-month plan establishing the organizational practices and tooling integrations that make maturity gains durable rather than temporary.

The Thirty-Day Quick Win Sprint

Every organization, regardless of its current maturity level, can produce observable improvements within 30 days without a new budget or headcount. For organizations at Level 1, the thirty-day sprint should establish infrastructure monitoring for every production AI endpoint: deploying an OpenTelemetry collector, configuring latency and error-rate metrics in an existing monitoring platform like Datadog or Grafana, and setting at least two alert thresholds with documented escalation paths. This is not Level 3 observability; it is the minimum viable foundation that prevents the most basic failure mode, an endpoint going dark for hours before anyone notices.

For organizations at Level 2, the thirty-day sprint should focus on establishing feature distribution baselines for the three highest-risk production models. Using Evidently or a custom Python script built on scipy, compute the training distribution for each key input feature, store those baselines in a version-controlled location, and write a scheduled job comparing current production feature distributions against those baselines weekly. The output does not need to be sophisticated; a simple PSI score logged to a monitoring dashboard is sufficient to demonstrate the concept and identify whether drift is occurring.

For organizations at Level 3 targeting Level 4, the thirty-day sprint should focus on one integration: connecting the observability platform's alert output to the enterprise ITSM system. Configuring a PagerDuty or ServiceNow webhook that fires when a high-priority drift alert triggers and routing it to the AI SRE on-call rotation transform model-quality alerts from dashboard observations into incidents that enter the standard response workflow. That single integration begins the transition to integrated operations, the defining characteristic of Level 4.

Diagram 14.7 — Thirty-Day Quick Win by Maturity Level

Thirty-Day Quick Win by Maturity Level.

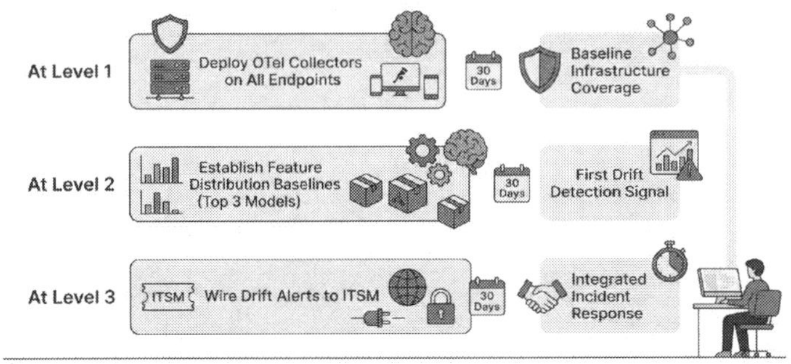

The Ninety-Day and Twelve-Month Roadmap

The ninety-day roadmap targets a single complete maturity-level transition. Moving from Level 2 to Level 3 in ninety days requires four parallel work streams: establishing model-specific performance metrics for all production models; deploying a drift detection tool with configured monitoring profiles; defining alert thresholds backed by statistical rationale; and conducting a model registry audit to ensure every production model has a documented baseline performance profile. Each work stream should have a named owner, a defined completion criterion, and a weekly check-in that surfaces blockers before they delay the overall timeline.

The twelve-month plan addresses durability. A maturity level is not achieved by completing a project; it is achieved when the practices required at that level are embedded in the organization's normal operating rhythm. At the twelve-month mark, capabilities gained in the ninety-day sprint should require no special management attention. The team should be running weekly drift review meetings as a standard agenda item. New models should arrive at the deployment gate with monitoring profiles pre-configured. Governance artifacts should be produced automatically by the deployment pipeline, not assembled by hand when an audit request arrives.

The twelve-month plan should include re-assessment using the self-assessment checklist at months three, six, and twelve. Each assessment produces a delta report: which capabilities have been fully operationalized, which are still partial, and what is blocking full completion. This cadence holds the team accountable for roadmap milestones. It provides the executive champion with evidence of progress to present to the board or regulators as part of the organization's AI governance narrative.

Diagram 14.8 — Ninety-Day to Twelve-Month Roadmap

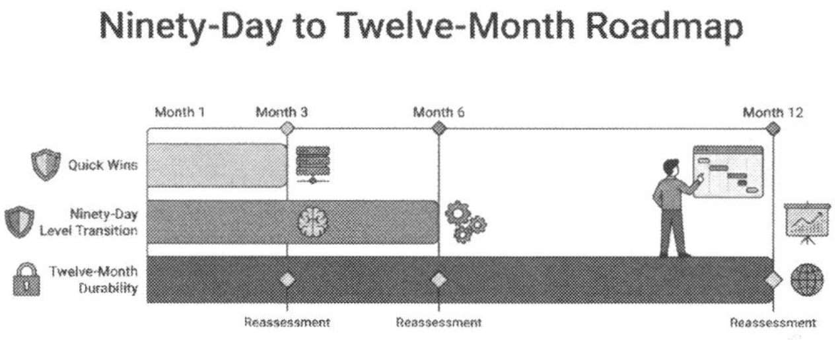

Ninety-Day to Twelve-Month Roadmap

Sustaining Maturity Gains Over Time

The most common failure mode in observability maturity programs is regression: organizations reach Level 3, experience a period of stability, reduce investment in the observability function, and slide back toward Level 2 as team members turn over, tools go unmaintained, and drift thresholds go unreviewed. Sustaining maturity gains requires three standing practices that are harder to establish than the initial technical implementation but are more important to long-term program health.

The first practice is a scheduled threshold review. Every alert threshold in the observability system has a staleness date. A PSI threshold calibrated eighteen months ago may be miscalibrated if business conditions have changed significantly. Quarterly threshold reviews, conducted by the Observability Lead in collaboration with the data science team, ensure that alert sensitivity remains appropriate. Thresholds that no longer reflect the current risk tolerance should be updated, and the rationale documented.

The organization should maintain a monthly record of which production models are fully covered by the monitoring program, which are partially covered, and which have no monitoring. This coverage report is the clearest indicator of whether the observability program is keeping pace with the portfolio. A portfolio that grows faster than observability coverage incrementally returns to Level 1 on new systems even as established systems remain at Level 3 or Level 4. An annual program review reassesses maturity level, evaluates the vendor landscape for new capabilities, and updates the roadmap for the coming year.

Diagram 14.9 — Maturity Sustainment Practices

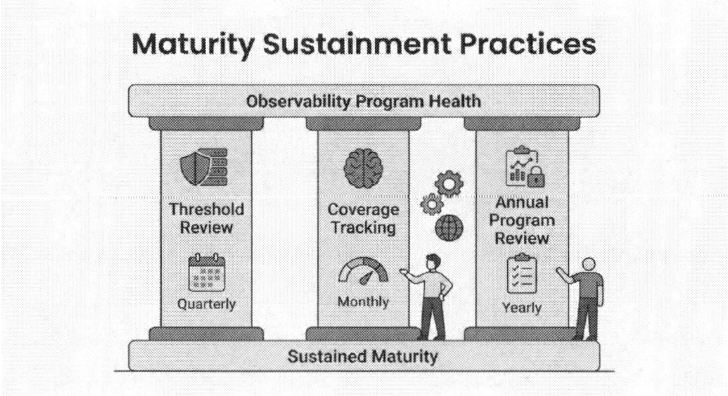

14.4 Manager's Checklist

- Conduct a self-assessment against the five-level maturity model using artifact-based evidence, not team assertions.
- Document the current maturity level for each production AI system separately; portfolio maturity is not uniform.
- Build a thirty-day quick win sprint targeting at least one concrete, measurable improvement at the current maturity level.
- Build a ninety-day roadmap targeting one complete level transition, with named owners and completion criteria for each work stream.
- Translate the gap analysis into a business case with three components: cost of current gap, investment required, and risk-adjusted return.
- Schedule quarterly threshold reviews for every alert policy in the observability platform and document the rationale for any changes.
- Maintain a monthly observability coverage report showing which production models are fully, partially, or not monitored.
- Conduct an annual program re-assessment using the self-assessment checklist and update the multi-year roadmap accordingly.

14.5 Takeaway

The AI observability maturity model converts the aspiration to improve AI monitoring into a specific, sequenced program with measurable milestones, artifact-based evidence of progress, and a financial case that can be defended in any budget conversation. The discipline it imposes, assessing capability against evidence rather than intention, distinguishes a production-grade AI program from a collection of pilots that happen to be running in production.

Maturity requires active maintenance through threshold reviews, coverage tracking, and periodic reassessment. Organizations that treat maturity gains as permanent without sustained investment slide back within eighteen months. The ones that sustain Level 4 and above treat observability as an operating practice, not a project, and hold their teams accountable to that distinction every quarter.

15 The Road Ahead: Observability in the Age of Autonomous AI

A major financial services firm deployed an autonomous AI agent to handle credit application intake. The agent used three tools: a credit bureau API, an internal fraud-scoring service, and a document-extraction pipeline. In its first week, the agent processed 4,200 applications without human intervention. Then it started failing silently. The fraud scoring service began returning elevated risk scores following an upstream model update, but the agent had no threshold logic to detect the shift. It began declining applications at 2.3 times the normal rate. The firm discovered the problem eleven days later when a portfolio analyst noticed an anomaly in approval volume. No traditional monitoring tool had flagged it: the agent's infrastructure metrics looked fine. Latency was normal. Error rates were zero. What the firm lacked was the ability to observe outcomes.

The enterprise AI landscape is changing faster than most observability programs are designed to keep pace. Autonomous agents that make sequences of decisions, call external tools, and collaborate with other agents are moving from experimental status into production. Edge AI deployments run inference on devices outside traditional monitoring infrastructure. Digital twin environments simulate the behavior of AI systems before changes are deployed. These shifts are reflected in current implementations across manufacturing, financial services, healthcare, and logistics. Observability programs designed only for batch prediction models and hosted LLM APIs will not cover them.

The manager's stake in this chapter is strategic planning. Decisions made today about observability architecture, vendor relationships, instrumentation standards, and team skills will determine whether the organization's monitoring program can extend to new deployment patterns or require a costly rebuild when they arrive in production. Understanding where the field is heading is a prerequisite for making those decisions well. This chapter maps the five most consequential shifts underway and argues for the specific management actions required for each.

15.1 From Monitoring Models to Observing Decisions

The transition from monitoring models to observing decisions is the most fundamental shift in AI observability practice over the next three to five years. Today's observability programs are largely model-centric: they track the inputs, outputs, and performance metrics of individual models. Tomorrow's programs must be outcome-centric: they must track the decisions that AI systems make, the downstream effects of those decisions on business outcomes, and the alignment between those decisions and the intent of the people who deployed the system.

Outcome Observability Defined

Outcome observability is the practice of monitoring the real-world effects of AI-generated decisions, not just the quality of the predictions that drive them. A credit scoring model may produce statistically stable predictions, but if the actual repayment behavior of approved applicants diverges from historical patterns, the model is not serving its intended purpose. Traditional model monitoring would not detect this divergence because it focuses on the predictive distribution rather than the outcome distribution. Outcome observability closes that gap by tracking downstream results of AI decisions over a time horizon long enough to observe actual consequences.

Implementing outcome observability requires two capabilities that most organizations do not yet have. The first is outcome linkage: a mechanism that traces each AI decision to the downstream event that constitutes its outcome, connecting a loan approval to a repayment record, an inventory reorder recommendation to actual stockout rates, and a healthcare triage score to actual length-of-stay. This linkage is a data engineering challenge because outcome data often lives in operational systems not designed to be joined with AI decision logs. The second capability is a time-delayed evaluation pipeline: a process that re-evaluates AI decisions retrospectively once outcome data is available.

For agentic AI systems, outcome observability extends beyond individual decisions to sequences of decisions. An autonomous procurement agent may execute twelve tool calls and three intermediate reasoning steps before generating a purchase order. Outcome observability for such a system requires session-level evaluation: the ability to assess whether the outcome of a complete agent session was aligned with the stated goal. This requires observability tooling that can represent and evaluate multi-step decision graphs, not just individual prediction events.

Diagram 15.1 — Model-Centric vs Outcome-Centric Observability

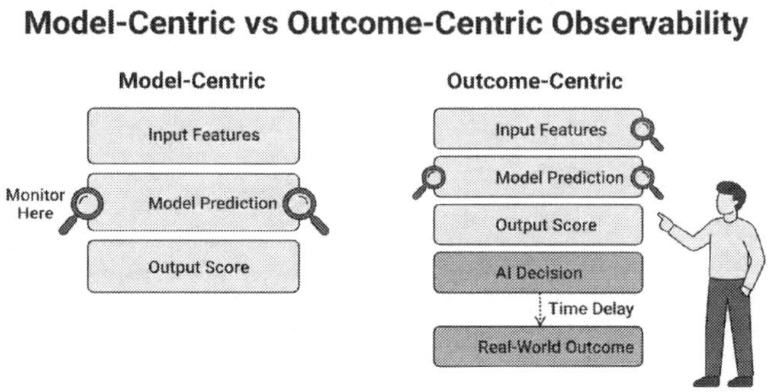

Observability for Agent-to-Agent Interactions

Multi-agent architectures introduce an observability challenge with no direct analog in traditional software monitoring. When a human-facing orchestrator agent delegates subtasks to a specialized coding agent, a document analysis agent, and a database query agent, the quality of the final output depends on the quality of each delegate's output. A failure in the document analysis agent may propagate silently to the orchestrator's response without triggering any error signal, because the delegate returned a response, just a wrong one. Detecting that class of failure requires evaluation at each delegation boundary, not just at the final output.

Implementing agent-to-agent observability requires trace propagation across agent boundaries. Each agent in a multi-agent system should emit spans that include the orchestrating session's parent trace ID, so that the complete decision graph from the initial user request through every delegation to the final response can be reconstructed and evaluated as a single trace. OpenTelemetry GenAI semantic conventions are beginning to define the attribute schemas needed to represent agent-to-agent communication in a standardized way. Organizations that implement against these conventions today will benefit from ecosystem tooling as it matures.

The governance question that agent-to-agent observability raises is data ownership. When does an orchestrator agent delegates to a third-party agent service, who owns the trace data from the delegate agent's execution? If the delegate agent processed personally identifiable information, that trace data may be subject to GDPR or CCPA retention and access controls. Managers should establish data ownership terms for all third-party agent services in procurement contracts, not as an afterthought during an audit inquiry.

Diagram 15.2 — Multi-Agent Trace Propagation Graph

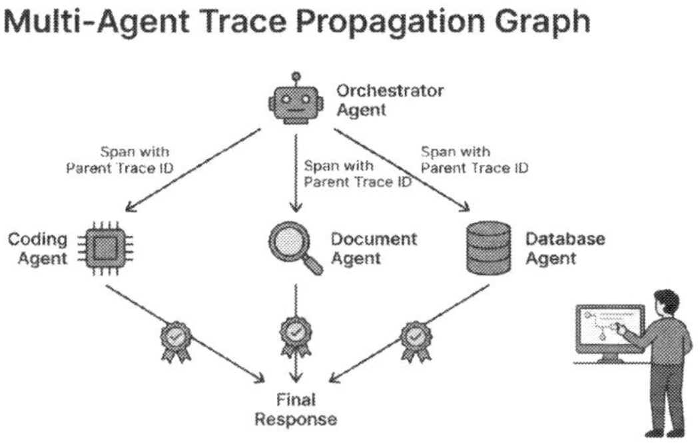

The Convergence of Observability, Security, and Governance

The operational boundaries between AI observability, AI security, and AI governance are dissolving. Prompt injection attacks are both security incidents and observability events. A bias-drift event in a high-risk model is both a governance concern and a model-quality issue.

Shadow model behavior, where a model produces outputs that meet statistical performance criteria but violate ethical constraints in edge cases, is simultaneously a model quality problem, a governance problem, and potentially a regulatory violation.

Vendors are responding by building unified platforms combining observability metrics with security monitoring and governance artifact generation in a single data pipeline. Arize is expanding from model performance monitoring into LLM security monitoring. Datadog and Dynatrace are adding AI-specific security modules. The implication is significant: organizations with separate point solutions may be maintaining three redundant data pipelines that produce overlapping signals, incurring substantial costs in money and operational complexity.

The manager's decision in this convergence is timing: when to consolidate separate observability, security, and governance tooling into a unified platform. The case for consolidation is compelling on cost and data-quality grounds; a single pipeline that logs, evaluates, and stores AI telemetry is cheaper to operate and yields richer, more correlated signals. The case for patience is that unified platforms are still maturing, and a premature platform migration that disrupts an established compliance audit trail can exceed the savings. A reasonable posture is to evaluate unified platforms over the next twelve months while designing new deployments to emit OpenTelemetry-compliant telemetry that any future platform can consume.

Diagram 15.3 — Convergence of Observability, Security, Governance

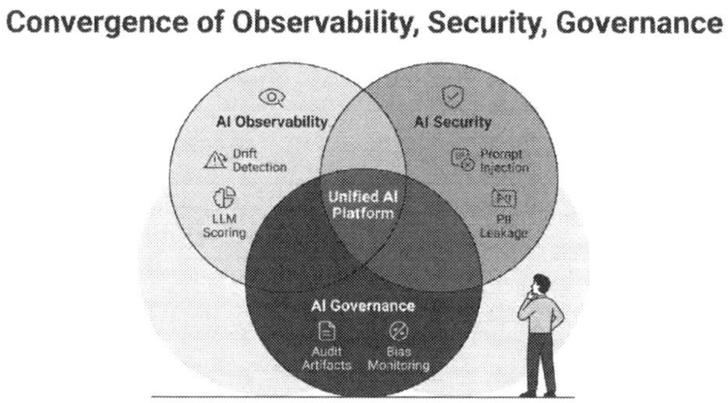

15.2 Edge AI, Digital Twins, and Emerging Standards

Three parallel developments are extending the boundaries of what AI observability must cover: inference models deployed on edge devices outside traditional network perimeters; digital twin environments for AI system simulation and pre-deployment testing; and emerging open standards that promise to standardize the telemetry emitted by AI systems. Each development creates new management decisions and new technical requirements.

Edge AI Observability

Edge AI observability monitors models running on devices outside the central data center: industrial sensors, autonomous vehicles, retail point-of-sale terminals, medical imaging devices, and smartphones. The fundamental challenge is connectivity: edge devices are often intermittently connected to the central network, which means telemetry cannot be streamed in real time. A model running on an autonomous guided vehicle in a warehouse may execute thousands of inferences between network synchronization cycles. Observability data must be buffered locally and transmitted in batches, introducing latency between inference execution and central monitoring visibility, ranging from seconds to hours.

The engineering approach requires lightweight telemetry agents that run on constrained hardware, local buffering with configurable flush intervals and overflow policies, and a central aggregation service that reconciles batch telemetry from thousands of devices. OpenTelemetry's Collector architecture, with its configurable pipeline of receivers, processors, and exporters, is well-suited to this pattern, and several edge computing platforms have begun shipping lightweight OTel Collector distributions optimized for embedded hardware.

The governance implications of edge AI observability are that the organization may have production AI systems making consequential decisions without real-time monitoring. A defect-detection model on a manufacturing line camera may be classifying parts incorrectly for four hours before the next telemetry batch synchronizes. The management response is to explicitly acknowledge observability gaps in risk documentation, establish local fallback behaviors when model confidence falls below a threshold, and define a maximum tolerable observability gap that constrains the telemetry synchronization frequency.

Diagram 15.4 — Edge AI Telemetry Pipeline

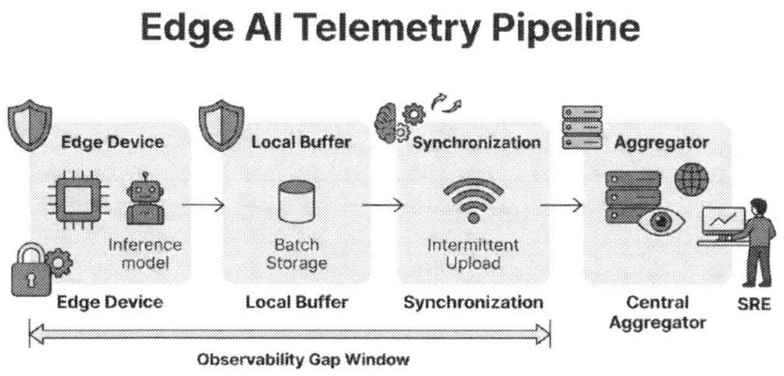

Digital Twins for AI System Simulation

In the AI context, a digital twin is a simulation environment that mirrors the data distribution, tool dependencies, and operational conditions of a production AI deployment, allowing teams to test model updates, new prompt templates, or agent configuration changes in realistic conditions before promoting to production. Their application to AI system observability is relatively recent and represents a significant advancement in maturity for pre-deployment risk management.

The most immediate value of a digital twin for AI observability is pre-deployment regression testing. When a production LLM application is updated with a new model version or revised system prompt, the digital twin can replay a representative sample of historical production traces against the new configuration and compare the output distribution to the baseline. If the new configuration had triggered safety guardrails, cost anomalies, or quality degradation signals on historical traffic, those issues would have been detected before affecting live users. This is shift-left observability: applying monitoring logic at development time rather than only at production time.

Maintaining a simulation environment that accurately reflects production data distributions requires ongoing effort as distributions evolve. The synthetic traffic generation or production replay mechanism must also be governed to avoid exposing real user data to a non-production environment. Organizations that have successfully deployed digital twin observability report the most valuable use cases are evaluating LLM prompt changes, which are notoriously difficult to regression-test without realistic traffic, and testing multi-agent workflow modifications, where component interactions make manual testing impractical.

Diagram 15.5 — Digital Twin Pre-Deployment Testing Flow

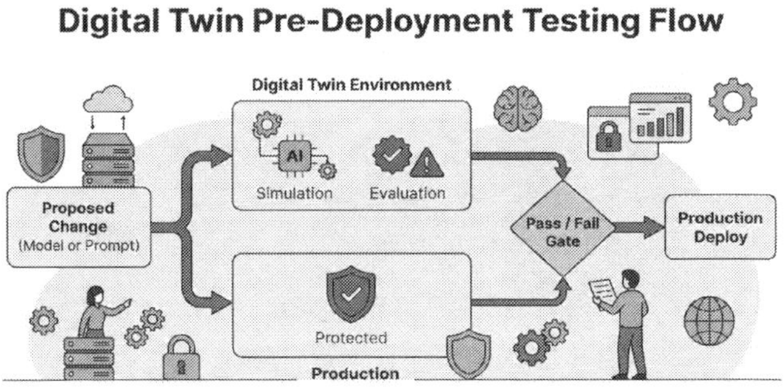

OpenTelemetry GenAI, OpenInference, and the Standards Landscape

The absence of standardized telemetry schemas has been a persistent friction point: every platform vendor defines its own data model for traces, spans, and evaluation metrics, making it difficult to switch vendors, correlate data across platforms, or build custom tooling that works consistently. Two open standards are beginning to address this fragmentation. OpenTelemetry GenAI defines standard attribute names and schemas for spans emitted by LLM calls, vector database queries, and AI model invocations. OpenInference, developed by Arize and increasingly adopted by other vendors, defines a similar schema optimized for LLM tracing and evaluation.

An organization that instruments its AI systems against OpenTelemetry GenAI semantic conventions creates telemetry that is portable across any observability backend supporting the standard, whether Phoenix, Datadog, or a custom platform. This reduces vendor lock-in at the data layer, the most important layer to keep portable, and positions the organization to benefit from ecosystem tooling, including open-source evaluation libraries and anomaly-detection models built against standard schemas.

The standards are still maturing, and coverage is uneven across AI system types. LLM call tracing is reasonably complete; semantic conventions for agent-to-agent communication, vector retrieval quality, and multi-modal inputs are still under active development. A pragmatic approach is to adopt the stable portions, specifically LLM span instrumentation and basic model invocation tracing, while maintaining flexibility for agent and multimodal workloads as the standards mature. Participating in the OpenTelemetry GenAI working group, even as a corporate observer, is a low-cost way to stay current on standards direction.

Diagram 15.6 — Open Standards Instrumentation Stack

Open Standards Instrumentation Stack

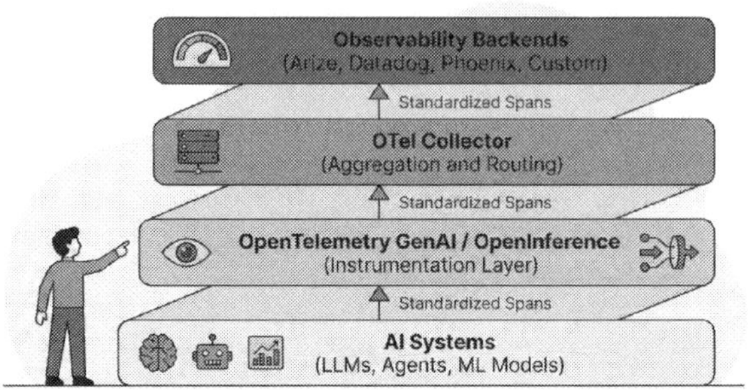

15.3 The Manager's Strategic Mandate

The technical shifts in this chapter share a common implication for enterprise management. AI observability is no longer a technical concern that can be delegated entirely to engineering teams. It is a strategic capability requiring management decisions about architecture, standards adoption, vendor relationships, team structure, and organizational accountability. This section translates those shifts into management actions that will determine whether the organization's observability program remains effective as the AI landscape evolves.

Making Observability a Strategic Capability

A strategic capability provides a durable competitive advantage, scales with organizational growth, and adapts to changing conditions without a complete rebuild. AI observability becomes strategic when designed to cover not just the current portfolio but any AI system the organization might deploy over the next three to five years. That design horizon requires deliberate choices: an instrumentation standard not tied to a single vendor; a data pipeline that ingests both batch and streaming telemetry;

and an evaluation framework that extends from traditional ML metrics to agent session metrics to outcome metrics.

The signal that observability has become strategic is when it appears in planning conversations that precede deployment decisions rather than remediation conversations that follow incidents. When the product team proposing a new autonomous agent deployment is asked, before the architecture review, how the agent's decisions will be monitored and what the escalation path will be for a decision quality event, observability has become a first-class concern. When the data science team proposes an edge model deployment that includes an observability gap analysis, observability has been internalized as an engineering discipline.

Reaching that state requires management action. It requires policy: an AI deployment standard that mandates observability design as a prerequisite for promotion to production. It requires accountability: an Observability Lead or VP of AI/ML who reviews every significant AI deployment proposal and signs off on the observability design. It requires investment: a platform that can scale to cover the full portfolio. And it requires measurement: a regular executive-level review of portfolio observability coverage, incident-detection performance, and program costs.

Diagram 15.7 — Observability as Strategic Capability

Observability as Strategic Capability

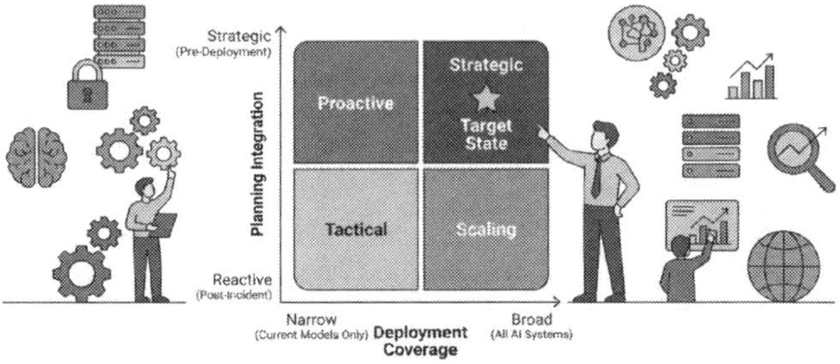

The Pilot-to-Production Pathway for New AI Paradigms

Each new AI deployment category creates its own pilot-to-production observability pathway challenge. The pattern from successful organizations is consistent: initiate the pilot from day one, not after it has demonstrated business value and promotion pressure has intensified. Instrumentation decisions made under promotion pressure are almost always insufficient because the team focuses on demonstrating the business case rather than designing for operational resilience.

For agentic AI pilots, the minimum instrumentation requirement from day one includes session-level trace collection for every agent run, logging of every tool call with input, output, and latency, and a baseline evaluation of final session outcomes against stated goals. This does not need to be automated in the pilot phase; even a structured human review of a sample of agent session traces each week provides enough signal to detect systematic failure patterns before they become production incidents. What it must be is consistent and documented, so that when the pilot is reviewed for production promotion, there is an actual performance record rather than an anecdote-driven assessment.

The pilot-to-production pathway for edge AI models should include an explicit assessment of observability gap risk in the production readiness review. That assessment should answer four questions: what is the maximum tolerable degradation period without central visibility; what local fallback behavior is triggered when model confidence falls below a defined threshold; what is the telemetry synchronization frequency and what happens if the device loses connectivity beyond the buffer capacity; and who is on call for edge model incidents and what is the escalation path when a batch synchronization reveals extended degradation.

Diagram 15.8 — Pilot-to-Production Observability Gate

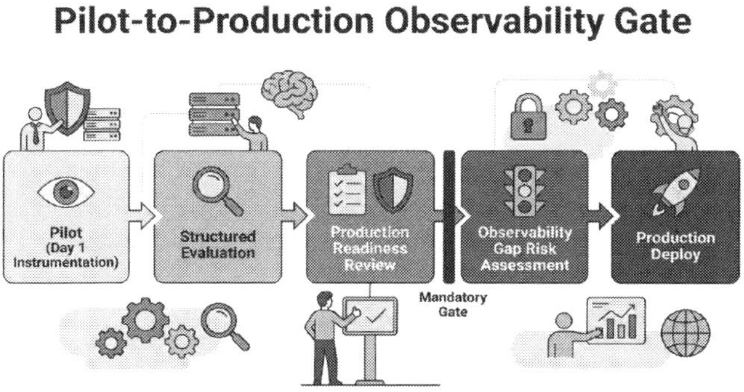

The Manager's Mandate for the Next Three Years

The three-year management mandate for AI observability comprises four commitments. The first is architectural commitment: adopt OpenTelemetry GenAI as the instrumentation standard for all new AI deployments and require existing deployments to migrate within eighteen months. This enables data portability, allowing vendor flexibility and ecosystem tooling adoption without a forklift migration whenever the observability platform evolves.

The second commitment is coverage: establish and maintain an observability coverage target, specifically the percentage of production AI systems with at least Level 3 observability, reviewed at the executive level quarterly. A common starting target is 80 percent Level 3 coverage within twelve months. The third is outcome commitment: extend the top 10 highest-risk production AI systems to outcome observability within 18 months, with documented outcome-linkage designs and time-delayed evaluation pipelines.

The fourth commitment is governance integration: ensure that by the end of the three-year horizon, AI observability outputs are formally integrated into the organization's GRC program, with automated audit artifact generation, documented retention policies, and at least one completed regulatory inquiry that used observability data as primary evidence. These four commitments convert the abstract ambition of mission-aligned AI observability into a concrete, time-bound program with executive ownership and measurable outcomes.

Diagram 15.9 — Four Commitments Three-Year Mandate

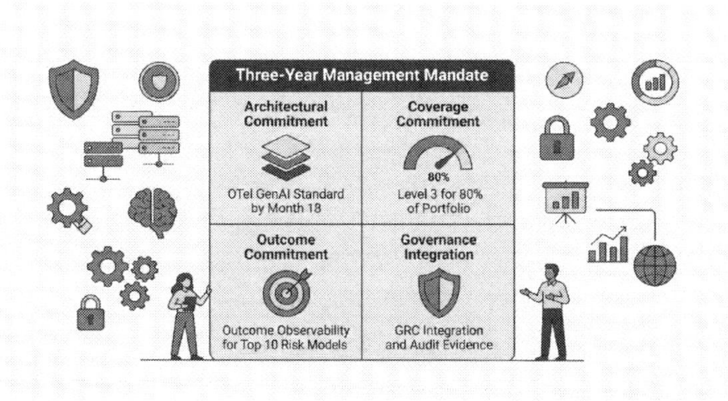

15.4 Manager's Checklist

- Require outcome linkage design as a mandatory artifact for every high-risk AI deployment, specifying how AI decisions will be traced to downstream business outcomes.
- Include data ownership terms for third-party agent services in all AI vendor procurement contracts, addressing trace data retention and access controls.
- Adopt OpenTelemetry GenAI semantic conventions as the instrumentation standard for all new AI deployments; set an eighteen-month migration target for existing systems.
- For every edge AI deployment, document the maximum tolerable observability gap, the local fallback behavior at low confidence, and the telemetry synchronization architecture.
- Establish a portfolio observability coverage target (percentage at Level 3 or above) and review it at the executive level quarterly.
- Require observability design review and sign-off before any autonomous agent or multi-agent system is promoted to production.

- Evaluate unified observability, security, and governance platforms in the next twelve months; design new deployments to emit OTel-compliant telemetry.
- Commission a digital twin evaluation for at least one high-stakes LLM application within twelve months, using historical production traces for pre-deployment regression testing.

15.5 Takeaway

The direction of AI observability is unambiguous: the scope of what must be observed is expanding from individual model outputs to agent decision sequences, real-world outcomes, edge-device inferences, and interactions between AI systems that increasingly operate without human intermediation at each step. The organizations that will govern this complexity effectively are not the ones with the most sophisticated tooling; they are the ones whose managers have made observability a first-class design requirement rather than a post-deployment remediation task.

The strategic mandate is straightforward to state, even if demanding to execute: instrument everything from day one, adopt open standards so you own your data regardless of which vendor you use, connect AI decisions to real outcomes so that mission-aligned performance is measurable rather than assumed, and hold the line on observability coverage as the portfolio grows. Every chapter in this book has been building toward that mandate. The road ahead will test whether organizations follow it.

16 Closing

The throughline of this book is a single, essential argument: AI observability is the operational discipline that makes enterprise AI trustworthy. Not trustworthy in an aspirational sense, where trust is asserted in a product announcement or a model card, but trustworthy in an operational sense, where trust is earned by demonstrating, continuously and systematically, that the AI systems your organization has deployed are still performing as intended, still operating within policy bounds, still generating the business value that justified the investment, and still aligned with the fairness and safety requirements that governance and regulatory frameworks demand. That kind of trust is not a property of a model at the moment it is deployed; it is a property of the monitoring, evaluation, and governance program that surrounds the model throughout its production life. The five layers of the observability stack, infrastructure health, data pipeline integrity, model performance, output evaluation, and agent decision tracing, are the operational instruments that make that demonstration possible. Build all five, connect them with open standards, and the result is an organization that can speak with authority about what its AI systems are doing and why. No observability program means no verifiable trust. No verifiable trust means no sustainable enterprise AI. The argument is direct.

You own the feedback loop. That is the central management principle this book has tried to establish at every level of its argument. The vendors who sell you infrastructure monitoring platforms, AI-native observability tools, LLM evaluation frameworks, and data pipeline quality platforms do not own the feedback loop. They provide the instruments. The data scientists who built the models do not own the feedback loop after deployment. They provided the initial intelligence. The ML platform engineers who manage the inference infrastructure do not own the feedback loop. They maintain the execution environment. You, the manager, own the feedback loop: the structured connection between what your AI systems are doing in production and the organizational decisions about whether to retrain, redeploy, restrict, or retire them. That connection is a management accountability, not a technical artifact. The tools, processes, roles, and review cadences described in this book are all in service of that accountability. When a drift alert fires on a credit-scoring model at two in the morning, the engineering team will know what happened and how. The question of what to do about it, whether to roll back, whether to notify the compliance office, whether to suspend the model pending investigation, whether to escalate to the board risk committee, is a management decision. The observability program provides the information; the manager makes the call. Owning the feedback loop means being ready to make that call.

The next five years will bring changes to AI observability that this book can only partially anticipate. Agentic AI systems, which make autonomous multi-step decisions across extended time horizons, will become the dominant deployment pattern in enterprise AI, and observability for decision graphs, tool call sequences, and multi-agent collaboration will move from an advanced capability to a baseline requirement. Every organization that is serious about autonomous AI will need an observability program that covers the full execution graph of every agent in production. That program will need to be governed with the same rigor that today's model risk management programs apply to predictive models.

Autonomous remediation will also mature: observability platforms will not only detect drift events but will initiate, and in some cases complete, corrective actions without human intervention, raising new governance questions about where human judgment must be preserved in the loop and what constitutes an acceptable level of AI autonomy in the remediation process. The regulatory landscape will converge toward explicit observability requirements; the EU AI Act's provisions for high-risk AI systems, the NIST AI Risk Management Framework's emphasis on continuous monitoring, and the Federal Reserve's SR 11-7 model risk management guidance are early chapters in a regulatory story that will eventually cover every significant enterprise AI deployment in every regulated industry. Organizations that have built mature observability programs now will find that regulatory compliance is an operational capability they already possess; organizations that have not will face an expensive and disruptive retrofit under external pressure. Outcome observability, the practice of monitoring not just model behavior but real-world decision quality and the downstream impact on people and processes, will emerge as the frontier of the discipline, connecting AI system monitoring to the ethical and social accountability frameworks that regulators, investors, and civil society are increasingly demanding.

The work this book describes is not finished when you implement the tools. It is a continuous discipline, much like the financial controls, security monitoring, and operational risk management that mature enterprises have embedded in their organizational fabric over decades. AI observability becomes a strategic capability when it is institutionalized: with named owners with clear accountability, a funded platform that scales with the AI portfolio, review cadences that link monitoring data to decision-making authority, and governance artifacts that satisfy regulatory scrutiny before the auditors arrive. Begin where you are. Implement the minimum viable monitoring for your highest-risk production model this quarter. Add the shared platform infrastructure in the next two quarters. Extend coverage systematically as the portfolio grows. Build the organizational roles and the cross-functional review process that make the program self-sustaining. In twelve to eighteen months, you will have transformed your organization's relationship to its own AI systems, from deploying and hoping to deploying and knowing. That transformation is not a technology project; it is a management commitment. This book has given you the framework. The decision to act is yours.

17 References

Accenture. *AI in Production: Monitoring, Drift, and Governance at Scale.* Accenture, 2024. (CH1)

Amazon Web Services. *Monitoring Machine Learning Models in Production.* AWS, 2023. (CH5)

Arrieta, Alejandro Barredo, et al. "Explainability and Monitoring in Modern AI Systems." *AI and Ethics*, vol. 4, 2024. (CH4)

Baylor, Dennis, et al. "Continuous Evaluation of ML Models in Production." *Google Research Papers*, 2023. (CH5)

Breck, Eric, et al. "The ML Test Score Revisited: Production Monitoring Lessons." *Proceedings of MLSys*, 2024. (CH1)

C3.ai Digital Transformation Institute. *AI Drift, Decay, and Reliability in Enterprise Systems.* C3.ai, 2023. (CH3)

Cisco. *Observability for AI-Driven Applications.* Cisco Systems, 2024. (CH4)

Cloud Security Alliance. *LLM Security and Monitoring Guidance.* CSA, 2024. (CH6)

Datadog. *AI Observability: Metrics, Traces, and Drift Detection.* Datadog, 2024. (CH4)

Deloitte. *AI Reliability and Monitoring in Regulated Industries.* Deloitte, 2023. (CH11)

Elastic. *End-to-End Observability for ML Pipelines.* Elastic, 2024. (CH4)

European Union Agency for Cybersecurity (ENISA). *AI Lifecycle Security and Monitoring.* ENISA, 2023. (CH11)

Fiddler AI. *State of AI Observability 2024.* Fiddler Labs, 2024. (CH2)

Forrester Research. *The Total Economic Impact of AI Observability Platforms.* Forrester, 2023. (CH1)

Google Cloud. *Best Practices for ML Monitoring and Drift Detection.* Google Cloud, 2024. (CH5)

IBM Research. *AI Governance and Observability for Enterprise Systems.* IBM, 2024. (CH11)

IDC. *AI Infrastructure and Observability Trends 2025.* IDC, 2025. (CH9)

KPMG. *Operationalizing AI Monitoring and Governance.* KPMG, 2024. (CH11)

LinkedIn Engineering. "Real-Time Drift Detection at Scale." *LinkedIn Engineering Papers*, 2023. (CH3)

Microsoft Azure. *Monitoring and Evaluating LLM Applications.* Microsoft, 2024. (CH6)

MIT Sloan Management Review. "Why AI Fails in Production—and How Observability Fixes It." *MIT SMR*, 2023. (CH1)

MLCommons. *AI System Evaluation and Continuous Benchmarking.* MLCommons, 2024. (CH4)

NVIDIA. *Observability for LLMs and Agentic Systems.* NVIDIA, 2024. (CH7)

OpenTelemetry Project. *Semantic Conventions for ML Observability.* CNCF, 2024. (CH4)

PwC. *AI Risk, Monitoring, and Post-Deployment Controls.* PwC, 2024. (CH11)

Salesforce Research. "Monitoring LLM Behavior in Enterprise Applications." *Salesforce AI Research*, 2024. (CH6)

SAS Institute. *Model Monitoring and Lifecycle Governance.* SAS, 2023. (CH5)

Seldon. *ML Observability in Production: 2024 Report.* Seldon, 2024. (CH2)

Snowflake. *Data Quality and Drift Monitoring for AI Pipelines.* Snowflake, 2024. (CH3)

Stanford HAI. *AI System Reliability and Real-World Performance.* Stanford University, 2024. (CH1)

Uber Engineering. "Michelangelo Observability: Monitoring ML at Scale." *Uber Engineering Papers*, 2023. (CH5)

World Economic Forum. *Responsible AI Deployment and Monitoring.* WEF, 2024. (CH11)

Zhang, Hanlin, et al. "Evaluating LLM Output Quality in Production." *Transactions on Machine Learning Research*, 2024. (CH6)

Zilliz. *Vector Databases and Observability for LLM Applications.* Zilliz Research, 2024. (CH6)

Made in the USA
Monee, IL
12 November 2026

faa7f314-63c7-4c6c-8567-f9ebbdd56878R01